ELEMENTARY

MECHANICAL DRAWING

FOR SCHOOL AND SHOP

BY

FRANK ABORN, B. S.

Drawing Master of Public Schools, Cleveland, Ohio

British Library Cataloguing-in-Publication Data
A catalogue record for this book is available from the
British Library

Technical Drawing and Drafting

Technical drawing, also known as 'drafting' or 'draughting', is the act and discipline of composing plans that visually communicate how something functions or is to be constructed.

It is essential for communicating ideas in industry, architecture and engineering. The need for precise communication in the preparation of a functional document distinguishes technical drawing from the expressive drawing of the visual arts. Whereas artistic drawings are subjectively interpreted, with multiply determined meanings, technical drawings generally have only one intended meaning. To make the drawings easier to understand, practitioners use familiar symbols, perspectives, units of measurement, notation systems, visual styles, and page layout. Together, such conventions constitute a visual language, and help to ensure that the drawing is unambiguous and relatively easy to understand.

There are many methods of constructing a technical drawing, and most simple among them is a sketch. A sketch is a quickly executed, freehand drawing that is not intended as a finished work. In general, sketching is a quick way to record an idea for later use, and architects sketches in particular (in a very similar manner to fine artists) serve as a way to try out different ideas and establish a composition before undertaking more finished work. Architects drawings can also be used to convince clients of the merits of a design, to enable a building constructor to use them, and as a record of completed work. In a similar manner to engineering (and all other technical drawings), there is a set of conventions (i.e particular views, measurements, scales, and cross-referencing) that are utilised.

As opposed to free-sketching, technical drawings usually utilise various manuals and instruments. The basic drafting procedure is to place a piece of paper (or other material) on a smooth surface with right-angle corners and straight sides – typically a drawing board. A sliding straightedge known as a 'T-square' is then placed on one of the sides, allowing it to be slid across the side of the table, and over the surface of the paper. Parallel lines can be drawn simply by moving the T-square and running a pencil along the edge, as well as holding devices such as set squares or triangles. Other tools can be used to draw curves and circles, and primary among these are the compasses, used for drawing simple arcs and circles. Drafting templates are also utilised in cases where the drafter has to create recurring objects in a drawing – a massive time-saving development.

This basic drafting system requires an accurate table and constant attention to the positioning of the tools. A common error is to allow the triangles to push the top of the T-square down slightly, thereby throwing off all the angles. Even tasks as simple as drawing two angled lines meeting at a point require a number of moves of the T-square and triangles, and in general drafting this can be a time consuming process. In addition to the mastery of the mechanics of drawing lines, arcs, circles (and text) onto a piece of paper – the drafting effort requires a thorough understanding of geometry, trigonometry and spatial comprehension. In all cases, it demands precision and accuracy, and attention to detail.

Conventionally, drawings were made in ink on paper or a similar material, and any copies required had to be laboriously made by hand. The twentieth century saw a shift to drawing on tracing paper, so that mechanical copies could be run off efficiently. This was a substantial development in the drafting process – only eclipsed in the twenty-first century with 'computer-aided-drawing' systems (CAD). Although classical draftsmen and women are still in high demand, the mechanics of the drafting task have largely been automated and accelerated through the use of such systems. The development of the computer had a major impact on the methods used to design and create technical drawings, making manual drawing almost obsolete, and opening up new possibilities of form using organic shapes and complex geometry.

Today, there are two types of computer-aided design systems used for the production of technical drawings; two dimensions ('2D') and three dimensions ('3D'). 2D CAD systems such as AutoCAD or MicroStation have largely replaced the paper drawing discipline. Lines, circles, arcs and curves are all created within the software. It is down to the technical drawing skill of the user to produce the drawing – though this method does allow for the making of numerous revisions, and modifications of original designs. 3D CAD systems such as Autodesk Inventor or SolidWorks first produce the geometry of the part, and the technical drawing comes from user defined views of the part. This means there is little scope for error once the parameters have been set.

Buildings, Aircraft, ships and cars are now all modelled, assembled and checked in 3D before technical drawings are released for manufacture.

Technical drawing is a skill that is essential for so many industries and endeavours, allowing complex ideas and designs to become reality. It is hoped that the current reader enjoys this book on the subject.

PREFACE.

In my endeavors to develop in the minds of pupils in grammar and high school classes a clear comprehension of the elementary principles of mechanical drawing, it was found that something more was needed than the copying of a few isolated problems into drawing-books. The student gained a better grasp of this subject by solving a limited number of written problems, illustrating a principle or method of mechanical drawing that had been previously explained by the teacher from the blackboard.

The problems thus accumulated have been arranged in the following pages with a view to economizing the teacher's time as well as to furnish the student with facilities for independent work.

To the Mechanic.

It has been my aim in the following pages to illustrate only one new principle or method of mechanical drawing in the same problem, and to make no more problems in any given case than are necessary to sufficiently illustrate the particular point under discussion. By so doing, I have been enabled to present short problems, and thereby to make each point of interest stand out in bold relief, and at the same time greatly to economize the time of the student. But this plan has prevented making the problems, as a whole, to apply to any particular trade or profession. This is no detriment to the work, however, although the individual who is seeking to learn that part of mechanical drawing which most especially applies to a particular trade may at first glance be disappointed, and think there is nothing here for him. To such let me say that the principles of mechanical drawing are every-where the same. Master these, and their application to ship-building, stair-building, architecture, sheet-metal work, etc., etc., is an easy matter.

FRANK ABORN.

CLEVELAND, O.,
July, 1886.

CONTENTS.

ELEMENTARY

MECHANICAL DRAWING.

DRAUGHTING TOOLS.

THE PENCIL.

For Geometrical and Constructive drawing, the pencil must be of such quality and hardness that it will take and hold a fine point.

The H grade Faber pencil is the best for school work.

Sharpening the Pencil.—For ruled lines, the pencil should be sharpened to a flat edge (Fig. 1), as with the common round point it is impossible to make the lines fine and of uniform width, which they must be if accuracy is to be secured.

For free-hand lines, nothing will do but a keen, common, round point.

As there are free-hand lines to be drawn, even in mechanical drawing, the student should have two pencils, one of which is sharpened to a flat, and the other to a round, point.

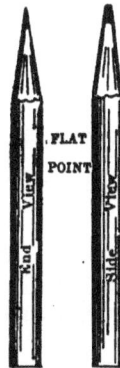

Fig. 1.

The Scale.

A **Scale** is a graduated ruler.

For school work, a common box-wood ruler, with beveled edge, and graduated to sixteenths of an inch, will serve the purpose very well.

Dividers.

There is no tool that is so irredeemably bad and entirely worthless as the dividers that are cheap imitations of the better class of instruments. They look well to the uninitiated, but the set screws do not fit, the joints are rough and soon grind themselves loose. Then, nothing will keep its place, and the instrument becomes unreliable and therefore useless. But the *Prang School Divider*, made of sheet metal, and costing twenty-five cents, is thoroughly reliable in every way for school work. If, however, one wishes to have something better than this, there is nothing in the market, that it is economy to buy, at a less cost than three dollars; and a set of instruments at this price

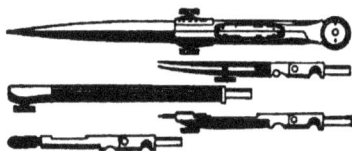

Fig. 2.

should consist only of a pair of dividers, with needle-point, pencil-point, pen, and lengthening-bar (Fig. 2); and any set of instruments consisting of a greater number of pieces than this must of necessity be correspondingly higher in price or poorer in quality.

T-Square.

A **T-Square** is an instrument, usually made of wood, consisting of a blade and head at right angles to each other (Fig. 3).

For school purposes, a T-square should have a blade not more than twelve inches in length.

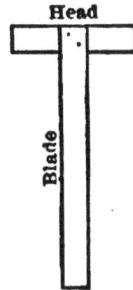

Fig. 3.

Fig. 4.   Fig. 5.

Set-Square.

A **Set-Square** is an instrument in the form of a right-angled triangle (Figs. 4 and 5).

Protractor.

A **Protractor** is an instrument usually in the form of a semicircular disk, having its periphery

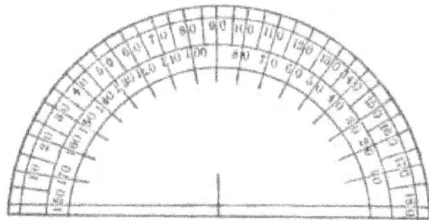

Fig. 6.

graduated into 180 degrees (Fig. 6).

For school work, a protractor made of horn, and three inches in diameter, is the best.

Summary.

Two H Faber pencils.
One pair Prang School Dividers.
One T-square, twelve inch blade.
One set-square, four inches on longest edge.

One box-wood foot rule, beveled edge, graduated to sixteenths of an inch.

One horn Protractor, three inches in diameter.

One slate, eight inches by ten inches; or, drawing board, ten inches by twelve inches, and suitable paper and thumb tacks.

PART I.

GEOMETRICAL DRAWING.

Geometrical Drawing is the description of lines arranged in conformity to some general rule called a geometrical law.

Lines are of two kinds,—straight and curved.

Straight Lines are horizontal, vertical, and oblique, according to their direction with reference to the plane of the earth's surface.

Curved Lines are either regular or irregular.

A curve is **Regular** when its curvature follows an established law.

A curve is **Irregular** when its curvature is not governed by any known rule or law.

CHAPTER I.—STRAIGHT LINES.

Section I.—Horizontal Lines.

All level lines, *i. e.*, lines parallel to the plane of the earth's surface, are *horizontal*.

Horizontal Lines are drawn with the help of the T-square.

Prob. 1.—Draw a horizontal line 4 in. long.

Fig. 7.   Fig. 8.

Solution.—Place the head of the T-square firmly against the left-hand edge of the slate or drawing board (Fig. 7,) and draw a line that is ·4 in. in length along the edge of the blade.

Prob. 2.—Draw a horizontal line 2½ in. long.

Prob. 3.—Draw four horizontal lines 3¼ in. long and ¼ in. apart.

Prob. 4.—Draw three horizontal lines, 1 in., 2 in., and 3 in. in length, and ⅜ in. apart.

Prob. 5.—Draw six horizontal lines, ⅛ in. apart and 2¼ in. long, with their ends in a plumb line.

Section II.—Vertical Lines.

All plumb lines, *i. e.*, lines that are perpendicular to the plane of the earth's surface, are *vertical*.

Vertical Lines are represented with the aid of the T-square and set-square.

Prob. 1.—Draw a vertical line 3½ in. long.

Solution.—Place the head of the T-square firmly against the left-hand edge of the slate or drawing board, and while in this position set one of the

shorter edges of the set-square against it (Fig. 8.)
Now, along the upright edge of the set-square, draw
a line 3½ in. long, which will be the line required
in the problem.

Prob. 2.—Draw a vertical line 4½ in. long.
Prob. 3.—Draw two vertical lines 2¾ in. long and
¾ in. apart.
Prob. 4.—Draw four vertical lines ½ in., 1 in., 2
in., and 3½ in. long, and ⅜ in. apart.

CHAPTER II.—The Circle.

Section I.—Definitions.

A **Circle** is a plane figure bounded by a curved
line, called its *circumference*, every part of which is
equally distant from a point within it called its *center*.

A **Diameter** of a circle is a straight line joining
two points in the circumference, and passing through
the center. Every circle may have an infinite number of diameters.

A **Radius** is a line extending from the center to
the circumference. It is
one half of a diameter.

Prob. 1.—Describe a circle $\frac{9}{32}$ in. in radius.

Solution.—Set the divid-
ers so that the distance
between the needle-point
and the pencil-point is $\frac{9}{32}$ in. Place the needle-

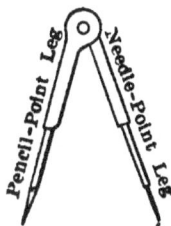

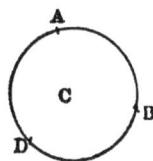

Fig. 9.   Fig. 10.

point in the position of the center *C,* and holding
the needle-point leg as nearly upright as possible,
revolve the pencil leg about it, so that the pencil
shall describe a continuous line *ABD,* every part of
which is equally distant from its center *C.*

Prob. 2.—Describe a circle 1 in. radius.
Prob. 3.—Describe a circle ½ in. radius.
Prob. 4.—Describe a circle 1¼ in. radius.
Prob. 5.—Describe a circle 1⅝ in. radius.
Prob. 6.—Describe a circle 1½ in. diameter.
Prob. 7.—Describe a circle 1 in. diameter.
Prob. 8.—Describe a circle 1¼ in. diameter.
Prob. 9.—Describe a circle 1⅝ in. diameter.
Prob. 10.—Describe a circle 1⅝ in. diameter.

Section II.—Arcs of Circles.

Any portion of a circumference less than the
whole is called an arc.

Every circumference is considered as consisting
of 360 equal arcs.

Each of these 360 arcs is called an arc of 1 degree.

The name of an arc depends upon the number of
degrees that it contains.

One fourth of a circumference is an arc of 90
degrees, and is written 90°.

One third of a circumference is an arc of 120
degrees, and is written 120°.

Three fourths of a circumference is an arc of 270
degrees, and is written 270°.

One half of a circumference is an arc of 180 degrees, and is written 180°, etc., etc.

A Protractor used in school work is usually a semicircular disk, and therefore its arc contains 180°. Usually these degrees are marked in two lines. One of these lines gives the number of degrees, counting from the

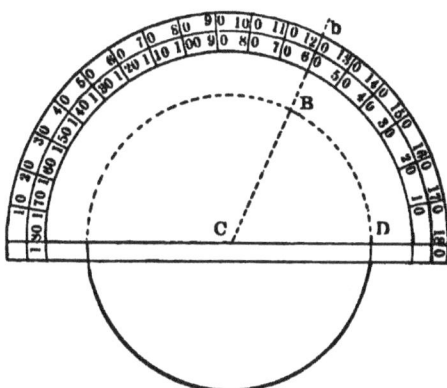

Fig. 11.

left-hand end of the diameter, and one gives the number of degrees counting from the right-hand end of the diameter.

Prob. 1.—Describe an arc of 60°, with a radius of ⅝ in.

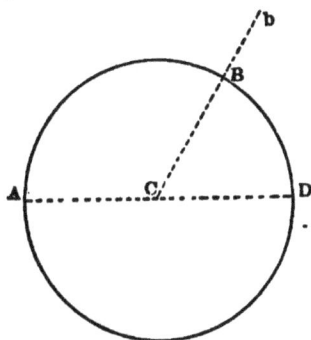

Fig. 12.

SOLUTION.—Describe a circle *ABD*, ⅝ in. in radius. Draw a diameter *AD*. Place the protractor on the circle *ABD*, so that its center and diameter coincide with the center and diameter of the circle, and mark the 60° point, *b*, at its edge. (Fig. 11.) Remove the protractor, and draw the line *bBC* to the center. (Fig. 12.) The point *B*, where the line crosses the circumference, will be one end of the required arc; the other end is at D, the end of the diameter from which it was measured. *BD* is the required arc.

Prob. 2.—Describe an arc of 120°.   Radius, $\frac{7}{8}$ in.

Prob. 3.—Describe an arc of 90°.   Radius, $1\frac{1}{4}$ in.

Prob. 4.—Describe an arc of 30°.   Radius, $1\frac{3}{8}$ in.

Prob. 5.—Describe an arc of 270°.   Radius, $1\frac{5}{16}$ in.

Prob. 6.—Describe an arc of 45°.   Radius, $\frac{9}{32}$ in.

Section III.—To describe an arc with a given radius that will be equal in length to the circumference of a given circle.

The length of a degree varies with the length of the radius of the circle.

Prob. 1.—With a radius of $\frac{7}{8}$ in., describe an arc that will be equal in length to the circumference of a circle having a radius of $\frac{3}{8}$ in.

Solution.—To find what part of the circumference of the larger circle is equal to the whole circumference of the smaller, we divide the radius of the smaller circle by the radius of the larger: $\frac{\frac{3}{8}}{\frac{7}{8}}=\frac{3}{7}$. Hence, an arc which is $\frac{3}{7}$ of the circumference of a circle whose radius is $\frac{7}{8}$ in. will be equal to the entire circumference of a circle having a radius of $\frac{3}{8}$ in. As there are 360° in an entire circumference, $\frac{1}{7}$ of it would be $\frac{1}{7}$ of 360° = 51.43°, and $\frac{3}{7}$ of it would be three times 51.43° =

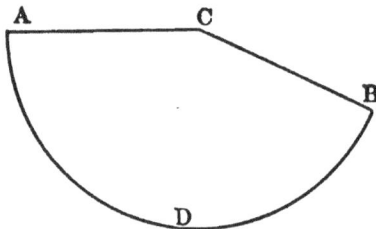

Fig. 13.

154.29°. Describe an arc, *ADB*, having a radius of ⅞ in., and on it lay off *AB* containing 154.29°. This arc will be equal in length to the circle having a radius of ⅝ in. (Fig. 13.)

Prob. 2.—With a radius of 2¼ in., describe an arc of a circle which is equal to the circumference of a circle of ¾ in. radius.

Prob. 3.—With a radius of 2¼ in., describe an arc of a circle which is equal to the circumference of a circle of ⅞ in. radius.

Prob. 4.—Describe an arc of a circle having a radius of 1¼ in., which is equal to the circumference of a circle, radius 1¾ in.

Prob. 5.—Describe an arc, radius 1⅜ in., equal to the circumference of a circle, radius 1¼ in.

Prob. 6.—Describe an arc, radius 1⅝ in., equal to the circumference of a circle, radius $\frac{5}{16}$ in.

Rule.—Divide the radius of the given circle by the radius of the required arc. Reduce the resulting fraction to degrees, and lay off the number of degrees thus found on the circumference of a circle having the radius of the required arc. The arc so laid off will be equal to the circumference of the given circle.

Section IV.—To find the Circumference of a Circle.

The **Circumference** of a circle is equal to the product of the diameter multiplied by 3.1416.

Note.—The exact ratio between the diameter and the circumference of a circle can not be given in figures, but 3.1416 is near enough for all ordinary purposes.

M. D.—2.

Prob. 1.—What is the length of the circumference of a circle that is 1½ in. in diameter?

SOLUTION.—As the diameter is 1.5 in., the length of the circumference must be 3.1416 times 1.5 in., which equals 4.7124 in. Hence, 4.7124 in. is the length of a circle 1½ in. in diameter.

Prob. 2.—What is the length of the circumference of a circle having a diameter of 2 in.?

Prob. 3.—What is the length of the circumference of a circle 4½ in. radius?

Prob. 4.—What is the diameter of a circle having a circumference of 6.2832 in.?

Prob. 5.—What is the length of the circumference of a circle having a radius of 3 in.?

Prob. 6.—What is the radius of a circle having a circumference of 10 in.?

Prob. 7.—Describe a circle having a circumference of 5.4978 in.

Prob. 8.—Describe a semi-circumference 2.7489 in. long.

Prob. 9.—Describe a quarter circumference 7.854 in. long.

Prob. 10.—Describe an arc of 45°, 5.8901 in. long.

Section V.—Bisection of Circular Arcs.

To bisect an arc is to divide it into two equal parts.

Prob. 1.—Bisect an arc of 60°; radius, 1¾ in.

SOLUTION.—Describe a circle $\frac{13}{32}$ in. in radius, and on its circumference lay off an arc of 60°, *AB*. With *A* and *B* as centers, describe arcs of equal radii intersecting in *b* and *b'*. Join *b* and *b'* by a straight line, and this line, where it cuts *AB* in *a*, bisects the arc *AB*. *Aa* and *aB* are equal.

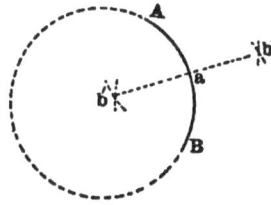

Fig. 14.

Prob. 2.—Bisect an arc of 90°, $1\frac{1}{4}$ in. radius.

Prob. 3.—Bisect a semi-circumference $\frac{3}{4}$ in. radius.

Prob. 4.—Divide an arc of 120° into four equal arcs. Diameter, $1\frac{7}{8}$ in.

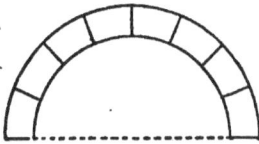

Fig. 15.

Prob. 5.—Divide an arc of 170°, $1\frac{1}{4}$ in. radius, into eight equal arcs.

Prob. 6.—Represent the face of a semicircular arch consisting of eight equal blocks. Radius of inner circle, $1\frac{1}{2}$ in., and radius of outer circle, 2 in. (Fig. 15.)

Prob. 7.—An arch is 4 in. in diameter on the inner circle. The key-stone is 20° wide and $\frac{7}{8}$ in. long measured on the radius, and on each side of the key are eight equal blocks $\frac{3}{4}$ in. long measured on the radii. Represent the face of the arch. (Fig. 16.)

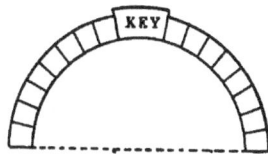

Fig. 16.

Section VI.—Trisection of Circles.

To trisect a circle is to divide it into three equal arcs.

A **Chord** of an arc is a line joining its extremities. (See Fig. 17.)

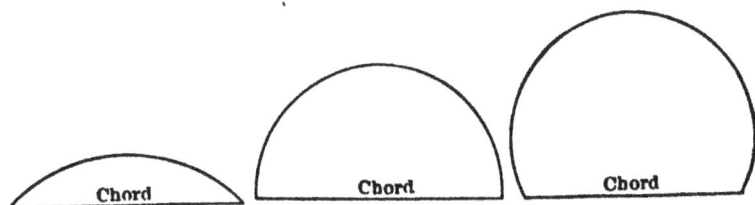

Fig. 17.

The chord of an arc of 60° is equal to the radius of the arc.

Prob. 1.—Draw a circle $\frac{3}{4}$ in. in diameter, and divide its circumference into six equal arcs.

SOLUTION.—Describe a circle AF, $\frac{3}{4}$ in. in diameter, and, without changing the dividers, set the needle-point at any point in the circumference, A, and describe arcs, cutting the circumference in B and F. Set the needle-point at F and B, and describe arcs cutting the circumference in E and C.

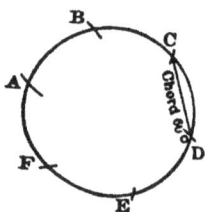

Fig. 18.

So proceed, and the last arcs will meet in D, and the circumference will be divided into six equal arcs. The chord of any of these arcs is equal to the radius of the circle.

To divide the circumference of a circle into three equal divisions: Divide the circumference into six equal arcs, and two of these arcs will constitute one third of the whole circumference.

To divide the circumference of a circle into twelve equal divisions: Divide the circumference into six equal arcs, and then divide each of these into two equal arcs.

Prob. 2.—Describe two circles, 1½ in. and ¾ in. in diameter, on the same center, and divide the annular space between them into twelve equal parts.

Prob. 3.—Describe a semi-circumference, diameter 2¾ in., and divide it into three equal arcs.

Prob. 4.—Describe a circle; divide its circumference into three equal arcs, and join the division points. The inscribed figure will be an equilateral triangle.

Prob. 5.—Describe a circle; divide its circumference into six equal arcs, and join the division points. The inscribed figure will be a hexagon (regular six sided polygon).

Section VII.—Concentric or Parallel Circles.

Circles are concentric when they have a common center: *a*, *b*, *d*, and *e* are concentric circles because they have the common center *C*.

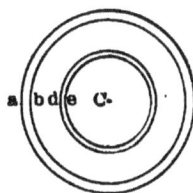

Fig. 19.

NOTE.—In describing circles, there is a strong temptation to incline the dividers to one side. In doing this, the center of the circle is so torn and enlarged that it can not be used again. It becomes necessary, therefore, for the student to practice making concentric circles until he has learned to hold the dividers *erect.*

Prob. 1.—Describe six concentric circles, ¼ in. apart, the radius of the largest circle being 2 in.

Prob. 2.—Describe six concentric circles, ½ in., 1 in., 1½ in., 2 in., 2½ in. in radius.

Prob. 3.—Describe three concentric circles, $3\frac{1}{4}$ in., $3\frac{1}{8}$ in., and 3 in. in diameter.

Prob. 4.—Describe four concentric circles, $\frac{7}{16}$ in., $\frac{9}{16}$ in., $\frac{10}{16}$ in., and $\frac{12}{16}$ in. in radius.

Prob. 5.—Represent the top of your ink well.

Prob. 6.—Describe two concentric circles, 2 in. and $2\frac{1}{4}$ in. in diameter. Divide the circumference of the smaller one into six equal parts, and with each of the division points as a center, describe six pairs of concentric circles, similar and equal to the first.

Section VIII.—Tangent Circles.

Circles are tangent when their circumferences touch each other in one point only. Circles may be tangent *externally* or *internally*.

Externally Tangent Circles.—Circles are externally tangent when the distance between their centers is equal to the sum of their radii.

Prob. 1.—Describe two circles externally tangent to each other; radii, $\frac{3}{16}$ in. and $\frac{9}{32}$ in.

SOLUTION.—As these circles are externally tangent, the distance between their centers will be equal to the sum of their radii: $\frac{3}{16}$ in.$+\frac{9}{32}$ in.$=\frac{15}{32}$ in. Draw a line CC'' $\frac{15}{32}$ in. long (Fig. 20.) With C as a center, and with a radius of $\frac{3}{16}$ in., describe the circle A; with C'' as a center,

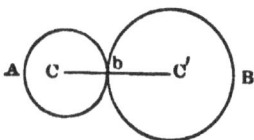

Fig. 20.

and with a radius of $\frac{9}{32}$ in., describe the circle B. These circles, A and B, will be externally tangent at b.

Prob. 2.—Describe two externally tangent circles; radii, ½ in. and ¾ in.

Prob. 3.—Describe two externally tangent circles; radii, $\frac{15}{16}$ in. and ⅜ in.

Prob. 4.—Describe two externally tangent circles; the sum of the radii to be 3 in., and one radius to be twice as long as the other.

Prob. 5.—Describe two concentric circles, ½ in. and ¾ in. in radius, externally tangent to two similar and equal concentric circles, with the larger of one pair of circles tangent to the smaller of the other pair.

Rule.—Draw a line equal in length to the sum of the given radii, and, with the ends of this line as centers, describe the given circles. The circles thus described will be externally tangent, and the point of contact will be on the line joining their centers.

Internally Tangent Circles.—Circles are internally tangent when the distance between their centers is equal to the difference of their radii.

Prob. 1.—Draw two internally tangent circles; radii, $\frac{3}{16}$ in and $\frac{9}{16}$ in.

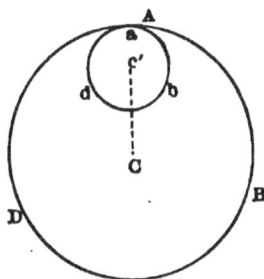

Fig. 21.

Solution.—Since $\frac{9}{16} - \frac{3}{16} = \frac{6}{16}$, the distance between the centers of the given circles is $\frac{6}{16}$ in. Draw a line, Cc' (Fig. 21,) $\frac{6}{16}$ in. in length, and with c' and C as centers, describe two circles, *abd* and *ABD*, $\frac{3}{16}$ in. and $\frac{9}{16}$ in. radii. *ABD* and *abd* are internally tangent at *a*.

Prob. 2.—Describe two internally tangent circles, 1 in. and ⅛ in. radii.

Prob. 3.—Describe two internally tangent circles, 1¾ in. and 1½ in. in diameter.

Prob. 4.—Describe two arcs of 90°, 9⁄16 in. and ⅞ in. radii, internally tangent at one end, and two other similar arcs similarly tangent, having a common point of contact, and all the centers on the same right line.

Rule.—Draw a line equal to the difference of the radii of the tangent circles, and with the ends of this line as centers describe circles with the given radii. The circles so drawn will be internally tangent to each other.

CHAPTER III.—ANGLES.

An angle is formed by the meeting of any two lines. (See Fig. 22.)

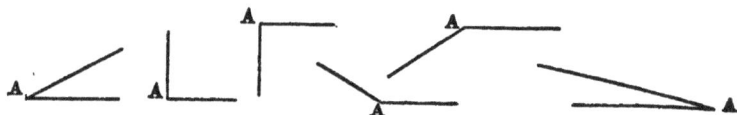

Fig. 22.

The point of meeting of two lines is called the apex of the angle.

If the apex of an angle be made the center of a circle, the arc intercepted by the sides of the angle is said to *subtend* the angle.

An angle is measured by the subtending arc. (See (Fig. 23).

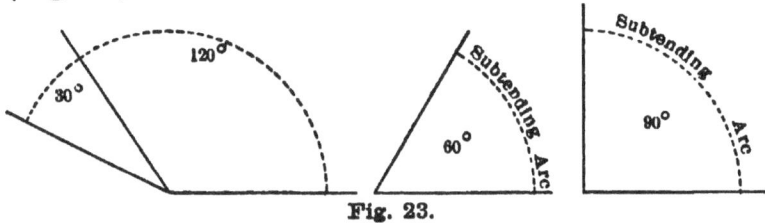

Fig. 23.

Prob. 1.—Draw an angle of 30°.

SOLUTION.—With any radius describe a circle *ABD* (Fig. 24). Lay off on this circle an arc of 30°, *AB*. From the center, *C*, draw radial lines through *A* and *B*. The arc *AB*, 30°, subtends and therefore measures the angle *ACB*.

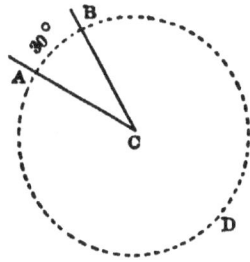

Fig. 24.

Prob. 2.—Draw two lines, 1 in. and 1¼ in. in length, forming an angle of 45°.

Prob. 3.—Draw three lines, 2 in. in length, meeting in a common point, and forming three angles of 120° each.

Prob. 4.—Draw three lines, 1½ in. in length, meeting in a common point, and forming three angles of 210°, 60°, and 90°.

Prob. 5.—Draw a horizontal line, 2 in. long, bisecting two lines 1½ in. in length, one of which makes an angle of 45°, and the other an angle of 135°, with the connecting line.

Rule.—Describe an arc containing the number of degrees in the required angle. Draw the radii at the ends of this arc. The angle formed by the radii thus drawn will be the angle required.

M. D.—3.

CHAPTER IV.—Triangles.

A **Polygon** is a plane figure having many sides and an equal number of angles.

A **Triangle** is a polygon that has three angles.

Section I.— To draw a triangle when its three sides are given.

Prob. 1.—Draw a triangle, the sides of which are $\frac{23}{32}$ in., $\frac{27}{32}$ in., and $\frac{7}{16}$ in. in length.

Solution.—Draw a line, AB, $\frac{23}{32}$ in. in length, for one of the sides of the required triangle. With one of the ends of this line, A, as a center, describe an arc, ac, $\frac{7}{16}$ in. in radius, and with the other end of the line, B, as a center, describe an arc $\frac{27}{32}$ in. radius, bd, cutting ac in C. Draw AC and BC, and the figure thus formed, ABC, is the required triangle, of which the sides are $\frac{23}{32}$ in., $\frac{27}{32}$ in., and $\frac{7}{16}$ in. (Fig. 25.)

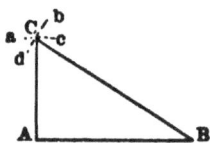
Fig. 25.

Prob. 2.—Draw a triangle, the sides of which are $1\frac{1}{8}$ in., $1\frac{1}{2}$ in., and 1 in. long.

Prob. 3.—Draw a triangle, the sides of which are 2 in., $\frac{3}{8}$ in., and $1\frac{5}{8}$ in. long.

Prob. 4.—Draw an equilateral triangle, the sides of which are $1\frac{7}{8}$ in. in length.

Rule.—Draw a line equal in length to one of the given sides of the required triangle, and, with the ends of this line as centers, describe intersecting arcs, the radii of which are equal to the other sides. From the point of intersection of these arcs, draw lines to the ends of the line already drawn. The three lines thus drawn will be the three sides of the required triangle.

Section II.—*To draw a triangle when one angle and two sides are given.*

Prob. 1.—Draw a triangle, of which one angle is 30°, and two of the sides are $\frac{9}{16}$ in. and $1\frac{3}{16}$ in. in length.

Solution.—Draw any line AB, $1\frac{3}{16}$ in. in length (Fig. 26.) At A draw a line AC, $\frac{9}{16}$ in. in length, and forming an angle of 30° with AB. Draw CB. The triangle ABC, thus formed, is the required triangle.

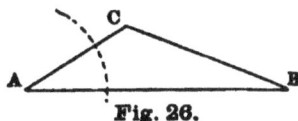
Fig. 26.

Prob. 2.—Draw a triangle, of which one angle is 45°, and two sides are 2 in. and $1\frac{3}{8}$ in. in length.

Prob. 3.—Draw a triangle, of which one angle is 90°, and two sides are 3 in. and $2\frac{1}{4}$ in. in length.

Prob. 4.—Draw a triangle, of which one angle is 60°, and two of the sides are $1\frac{5}{8}$ in. and $\frac{15}{16}$ in. in length.

Rule.—Draw two lines, equal in length to the given sides of the required triangle, and forming an angle equal to the given angle. Join the ends of these lines by a right line, and the three lines thus drawn will form the required triangle.

Section III.—To draw a triangle having given one side and two angles.

Prob. 1.—Draw a triangle, one side of which is $1\frac{5}{16}$ in. and two angles of which are 30° and 45°.

SOLUTION.—Draw a line, AB (Fig. 27), $1\frac{5}{16}$ in. in length, for one of the sides of the required triangle. At A draw AC, making an angle of 45° with AB, and at B draw BC, making an angle of 30° with AB.

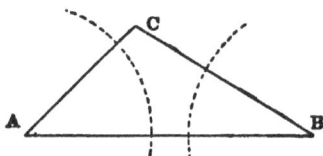

Fig. 27.

Produce these lines until they meet. Then ABC, having one side $1\frac{5}{16}$ in. in length, and two angles of 30° and 45° is the required triangle.

Prob. 2.—Draw a triangle, of which one side is $1\frac{1}{4}$ in. in length, and two angles are 45° and 90°.

Prob. 3.—Draw a triangle, of which one side is 1 in., and two angles are 60° each.

Prob. 4.—Draw a triangle, of which one side is $2\frac{1}{4}$ in. in length, and two angles are 30° and $22\frac{1}{2}$°.

Prob. 5.—Draw a triangle, of which one side is $\frac{5}{8}$ in. in length, and two angles are 75° each.

Rule.—Draw a line equal in length to the given side of the required triangle. At the ends of this line, and on the same side, draw lines forming the given angles of the required triangle. Produce these lines until they meet. The figure formed by the three lines thus drawn will be the required triangle.

CHAPTER V.—Perpendicular Lines.

Two lines are perpendicular to each other when the angles at their point of meeting are each 90°.

Section I.—To draw a line perpendicular to another line at a point in the middle of the given line.

Prob. 1.—Draw a line 1 in. long, and erect a perpendicular to it at a point midway between the ends.

Solution.—Draw a line, *AB* (Fig. 28), 1 in. in length. With *A* and *B* as centers, and the same radius, describe two arcs of circles, cutting each other in *b* and *b′* on both sides of the line *AB*. Join *b* and *b′*, and the line *bb′* will be perpendicular to the line *AB* and cut it in the middle.

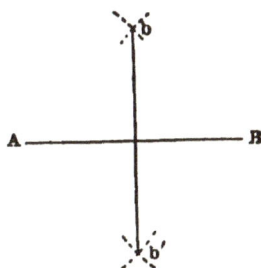

Fig. 28.

Prob. 2.—Draw a line 2¾ in. in length, and cross it in the middle by a line perpendicular to it. The cross line to be 1¾ in. in length and equally divided by the first line.

Rule.—Draw a line equal to the given line, and with the ends of this line as centers, describe two arcs, with the same radius, cutting each other on both sides of the given line. Join the points of intersection of these arcs, and the line thus drawn will be perpendicular to the given line at a point in the middle.

*Section II.—To draw a line perpendicular to another
at a point not in the middle of the given line.*

Prob. 1.—Draw a line 1⅛ in. in length, cut by a
line perpendicular to it at a point ½ inch from the
right-hand end of the given line.

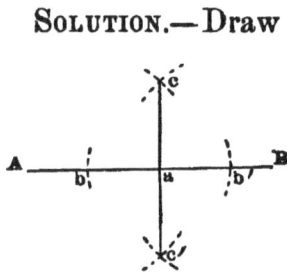

SOLUTION.—Draw a line *AB* (Fig. 29), 1⅛ in. in
length. Locate a point, *a*,
1 in. from its right-hand end.
With *a* as a center, describe
arcs of equal radii, cutting
AB in *b* and *b'*. With *b* and
b' as centers, describe arcs of
equal radii cutting each other

Fig. 29.

on both sides of *AB*, in *c* and *c'*. Join *c* and *c'*, and
the line *cc'* is perpendicular to *AB* at a point *a*, ½ in.
from the right-hand end of *AB*.

Prob. 2.—A line 1¾ in. in length is cut by another
line perpendicular to it at a point ⅞ in. from the
left-hand end. Draw the lines.

Prob. 3.—Draw a line 3 in. long, crossed 1⅜ in.
from its left-hand end by a perpendicular line 2 in.
long, which it bisects.

Rule.—Draw a line equal to the given line, and
in it locate the given point. With this point as a
center, describe arcs of equal radii cutting the
given line on each side of the given point. With
these points as centers, describe arcs of equal radii
cutting each other on both sides of the given line.

A line joining the points of intersection of these arcs will pass through the given point, perpendicular to the given line.

Section III.—To draw a line perpendicular to a given line from a point not in the given line.

Prob. 1.—Draw a triangle, the sides of which are $1\frac{7}{16}$ in., $1\frac{1}{16}$ in., and $\frac{5}{8}$ in. long, having a line perpendicular to its longest side drawn from the angle opposite it.

SOLUTION.—Draw the given triangle, *ABC*. (Fig. 30). With *C* as a center, describe an arc cutting the longest side, *AB*, in two points, *b* and *b'*. With these points as centers, describe arcs of equal radii cutting each other on both sides of *AB*, in *c* and *c'*. From *C* draw a line through *c* and *c'*. This line will be the required perpendicular.

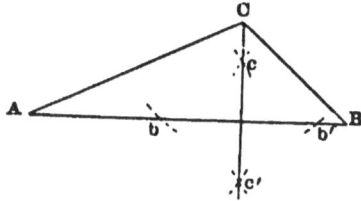

Fig. 30.

Prob. 2.—Draw a line $1\frac{3}{8}$ in. long, perpendicular to another line 2 in. long, from a point that is $1\frac{1}{2}$ in. from the left-hand, and $1\frac{7}{8}$ in. from the right-hand, end of the given line.

Rule.—Draw the given line, and locate the given point. With this point as a center, describe an arc cutting the given line in two points. With these points as centers, describe arcs of equal radii cutting each other on both sides of the given line. From the given point, draw a line through the points of intersection of these arcs, and this line will be the required perpendicular.

Section IV.—To draw a line perpendicular to, and at the end of, a given line.

Prob. 1.—The sides of a right-angled triangle adjacent to the right angle are 1 in. and ⅛ in. in length. Draw the triangle.

SOLUTION.—Draw a line *AB* (Fig. 31), 1 in. in length. With *B* as a center, describe a semi-circle, *bced*. Divide this semicircle into three equal arcs, *bc*, *ce*, and *ed*. Bisect the middle arc, *ce*, in *f*, and draw

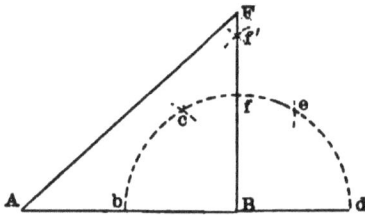
Fig. 31.

Bf. Extend *Bf* to *F*, making *BF* ⅛ in. in length. Join *A* and *F*, and *FAB* is the required right-angled triangle.

Prob. 2.—Draw a square; sides, 1¼ in. in length.
Prob. 3.—Draw a square; sides, 1⁵⁄₁₆ in., making angles of 30° and 60° with the horizon.
Prob. 4.—Draw a rectangle; sides, 1¼ in. and 1⅛ in. in length.

Rule.—Draw the given line, and, with one end of this line as a center, describe a semicircle, of which the given side produced is the diameter. Divide this semicircle into three equal arcs. Bisect the middle one of these arcs, and through the bisecting point draw a line to the end of the given line. This line will be the required perpendicular.

CHAPTER VI.—Tangent Lines.

A line is **Tangent** to a circle when it touches it at one point only.

The point where the line meets the circle is called the point of *tangency* or point of *contact*.

A tangent line is always perpendicular to the radius at the point of contact.

Prob. 1.—Draw a circle ½ in. in diameter, and a line 1 in. long tangent to the circle; the point of contact to be in the middle of the line.

Solution.—Draw the circle *AB* (Fig. 32), ½ in. in diameter. Draw the radius *CA*.
Draw the line *DE*, perpendicular to *CA*, at the end of it, and ½ in. on each side of the point of contact, *A*.
AB is the required circle, and *DE* is the required tangent to the circle.

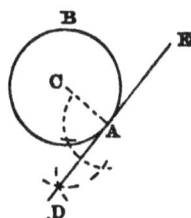
Fig. 32.

Prob. 2.—Draw a circle 1⅜ in. in diameter, and a tangent line 1 in. long; point of contact in the middle of the line.

Prob. 3.—Draw a circle ⅞ in. radius, and a tangent line 1¼ in. long; point of contact at the end of the tangent.

Prob. 4.—Describe two concentric circles ¾ in. and 1¼ in. diameter, and draw a tangent line to each of them 2 in. in length; points of contact on the same radius, and ¼ in. from the ends of the tangents.

Prob. 5.—Describe two circles 1 in. and ½ in. in diameter, 2 in. between centers, and one line tangent to both of them.

Rule.—Describe the given circle, and draw the radius at the point of tangency. Erect a perpendicular at the end of this radius, and the line so drawn will be tangent to the circle.

CHAPTER VII.—Parallel Lines.

Two lines are **Parallel** when the distance between them remains the same throughout their entire length.

Prob. 1.—Draw two parallel lines 1⅛ in. in length and ¼ in. apart.

Solution.—Draw a line AB (Fig. 33), 1¼ in. in length. As the parallel line is ¼ in. from this line, with any two points on the line AB, as centers, describe two arcs of ¼ in. radius on the same side of AB. Draw a line CD, 1⅛ in. long, tangent to these arcs, and this line will be parallel to AB and ¼ in. from it.

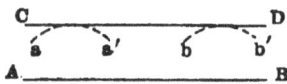

Fig. 33.

Prob. 2.—Draw three pairs of parallel lines : 3 in. long and ½ in. apart ; 5 in. long and ⅜ in. apart ; and 4¼ in. long and ⅝ in. apart.

Prob. 3.—Draw three sets of three′ parallel lines each : 3¼ in. long, ¼ in. apart ; 5¼ in. long, ⅛ in. apart ; 4⅛ in. long, ₁₆⁵ in. apart.

Prob. 4.—Draw three sets of four parallel lines : 3 in. long, ₁₆³ in. apart ; 4½ in. long, ¼ in. apart ; 5 in. long, ₁₆⁷ in. apart.

Prob. 5.—Draw the following described lines : first line, 1⅝ in. long ; second line, 2⅜ in. long, parallel with first line, and ⅝ in. from it ; third line, 3¼ in. long, parallel with second line, and ₁₆⁵ in. from it ; fourth line, 3⅝ in. long, parallel with third line, and ₁₆⁴ in. from it ; fifth line, 4⅝ in. long, parallel with fourth line, and ₁₆³ in. from it.

Rule.—Draw the given line, and, with two points in this line as centers, describe two arcs of equal radii on the same side of it. Draw a line tangent to these arcs, and this line will be parallel with the given line.

CHAPTER VIII.—To Divide a Right Line into any Number of Equal Parts.

Prob. 1.—Divide a line 1₁₆¹ in. long into five equal parts.

SOLUTION.—Draw a line 0-0 (Fig. 34), $1\frac{1}{8}$ in. long.

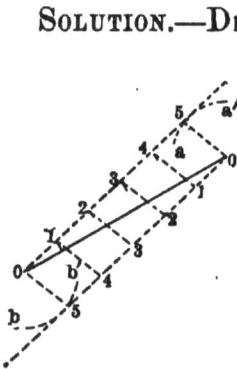

Fig. 34.

With 0 and 0 as centers, draw arcs *aa'* and *bb'* with equal radii on opposite sides of 0-0. From 0 draw the two lines 0-5 tangent to these arcs. Beginning at 0, lay off five equal divisions of any length on each of these lines, and join 0-5, 1-4, 2-3, 4-1, and 5-0. These lines will divide the given line, 0-0, into five equal parts.

Prob. 2.—Divide a line $4\frac{7}{16}$ in. long into four equal parts.

Prob. 3.—Divide a line $3\frac{5}{16}$ in. long into seven equal parts.

Prob. 4.—Divide a line $2\frac{11}{16}$ in. long into eleven equal parts.

Prob. 5.—Draw a line 3 in. long, and divide it into nine equal parts.

Prob. 6.—Draw a $4\frac{1}{2}$ in. square, and divide it into nine smaller squares.

Prob. 7.—Draw a circle $3\frac{7}{8}$ in. in diameter, crossed by four parallel lines perpendicular to its vertical diameter, and dividing it into five equal parts.

Prob. 8.—The sides of a triangle are $\frac{7}{8}$ in., $1\frac{1}{8}$ in., and $2\frac{1}{2}$ in., and it is crossed by three lines parallel with its longest side and dividing the other sides into four equal parts. Draw the triangle and the lines crossing it.

Rule.—Draw the given line, and with the ends of this line as centers, describe arcs of equal radii on opposite sides of the given line. From the ends of this line, draw two lines tangent to these arcs, and lay off on them the required number of equal divisions, beginning at the end of the given line. Join the corresponding division points, and the lines so drawn will divide the given line into the required number of equal parts.

CHAPTER IX.—THE ELLIPSE.

An **Ellipse** is a curved line that has two centers, called its foci, and two diameters, called its major and minor axes.

The **Major Axis** is the longest diameter of the ellipse.

The **Minor Axis** is the shortest diameter of the ellipse.

The major and minor axes of the ellipse are perpendicular to each other.

The sum of the distances of any point on the curve from the foci is equal to the major axis of the ellipse.

To Draw an Ellipse.—First Method.

Prob. 1.—Draw an ellipse, the axes of which are ⅛ in. and 2 in.

SOLUTION.—For the major and minor axes of the required ellipse, draw AB and DC, 2 in. and ⅛ in. in length, perpendicular to and bisecting each other. With the center at C, and the radius equal to one half of AB, describe arcs cutting AB in F and F'.

These points, F and F', are the foci of the required ellipse. To find a point through which an ellipse drawn on these diameters would pass, locate any point, E, on AB, and with AE and EB as radii, describe arcs from each of the foci as centers, cutting each other in e, e, e, and e.

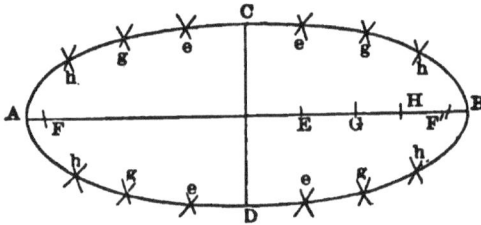

Fig. 35.

Then, in each case $eF + eF' = AB$, the major axis. In this way take as many points as necessary on AB, and proceed as before. When a sufficient number of points has been thus established, draw a curved line through them. This line will describe the required ellipse, *Aeg, Cge, Beg, Dge.* (Fig. 35.)

Prob. 2.—Draw an ellipse, of which the axes are 1½ in. and 2½ in.

Prob. 3.—Draw an ellipse, of which the axes are ⅜ in. and 1⅛ in.

Prob. 4.—Draw an ellipse, of which the axes are ½ in. and 1¾ in., and another on the same axes 1 in. × 2¼ in.

Prob. 5.—Draw two ellipses on the same diameters: ½ in. × 1¾ in. and 1 in. × 3½ in.

Prob. 6.—Draw an ellipse 1¼ in. × 2 in.

Prob. 7.—Draw two ellipses on the same diameters. The minor axis of one of them is ⅞ in., and of the other 1¾ in. The major axis in each is twice as great as the minor axis.

Rule.—Draw the major and the minor axes. Locate the foci. Divide the major axis into two unequal divisions. With each of the parts of the major axis thus obtained as radii, and the foci as centers, describe arcs intersecting in each quadrant of the ellipse. Make as many divisions of the major axis, and proceed as before, as may be necessary to determine the curve sufficiently. When all the points have been determined, draw a line joining them and the ends of the major and minor axes. The line so described will be the required ellipse.

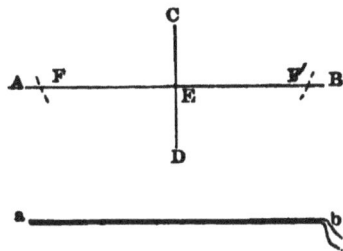

Second Method.—To Describe an Ellipse with a String.

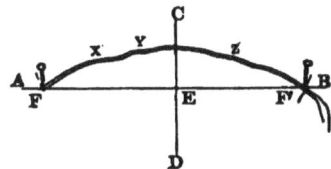

Note.—To describe an ellipse with a string is frequently very convenient; but as a string is elastic, it is unreliable. For this reason, it should never be used in describing ellipses, except where only quick, approximate results are required.

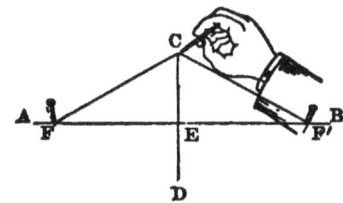

Prob. 1.—Draw an ellipse $\frac{9}{16}$ in. × $1\frac{3}{8}$ in.

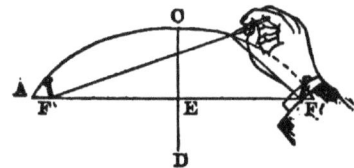

Solution. — Draw the axes of the ellipse, *AB*

Fig. 36.

and *CD* (Fig. 36). With *C* as a center, and *AE* as a radius, locate the foci *F* and *F'*. Double a string and tie a loop, the length of which, *ab*, is equal to *AB*. Set a pin in each focus, *F* and *F'*, and drop the loop over them, *xyz*. Place the pencil against the loop, and push it out to *C*. Then move the pencil either way, pressing against the string, and describe one half of the ellipse. In the same way, push the loop to *D* and repeat. The two lines so described will form an approximate ellipse.

Prob. 2.—Draw an ellipse; axes, 1½ in. and 2¼ in.

Prob. 3.—Draw an ellipse; axes, 1 in. and 1½ in.

Prob. 4.—Draw an ellipse; axes, 1½ in. and 1¾ in.

Prob. 5.—Draw an ellipse; axes, ½ in. and 2 in.

PART II.

CONSTRUCTIVE DRAWING.

INTRODUCTION.

Art aims so to present ideas as to elevate and develop sentiment and feeling. In other words, art is emotional, and therefore *variable*. Hence, any work may be excellent art, and still it may not be exactly and impartially true.

Science is matter-of-fact, practical, and therefore *exact*. It aims to state or describe only facts in the simplest and most unmistakable way.

All solids—objects that occupy space—have size; that is, every part of a solid has its own length, breadth, and height, called its dimensions.

The **Dimensions** of a solid may be described by two very different kinds of drawing, Pictorial Drawing and Constructive Drawing.

Pictorial Drawing is the art of describing the *appearance* of the dimensions of a solid on a flat surface.

Constructive Drawing is the science of describing the *exact dimensions* of a solid on a flat surface.

Every science has its exact and invariable methods, and the method of ascertaining the dimensions of a solid that is to be described by constructive drawing may be explained as follows:

On a common business card draw two lines, *ab*

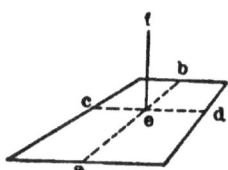

Fig. 37.

and *cd* (Fig. 37), perpendicular to each other. Through their point of intersection, *e*, thrust a straight common pin perpendicular to *ab* and *cd*. When the card lies on some level surface, so that *ab* and *cd* are horizontal, its length and breadth at *a* and *c* are measured on *ab* and *cd*, and the height or thickness is measured on the line *ef*.

If the card should stand so that *ef* and *cd* are horizontal (Fig. 38), the length would be measured on *cd*, and the breadth or thickness on *fe*, and the height on *ab*.

When the card stands so that *ef* and *ab* are horizontal (Fig. 39), the length and breadth or thickness are measured on the lines *ef* and *ab*, and the height is measured on the line *cd*.

Fig. 38.

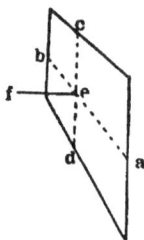

Fig. 39.

If the card should be inclined, and only *cd* remain horizontal (Fig. 39*a*), perhaps like the side of a gable roof, then the length would be measured on *cd*; but as *ef* would not be horizontal, nor

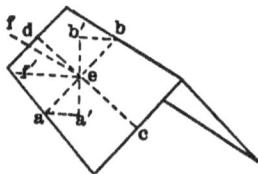

Fig. 39, *a*.

ab vertical, neither the height nor the breadth

could be measured on these lines, but on lines that are horizontal and vertical, as *ef'* and *a'eb'*.

From what precedes, it will be seen that whatever the form or position of the object to be described by *Constructive Drawing*, each part must be first measured; that is, the length and the breadth must be measured horizontally and perpendicular to each other; and the height must be measured vertically, perpendicular to both the length and the breadth.

For the exact description of these dimensions, *two different drawings are necessary*.

One of these drawings is always taken horizontally, parallel to the length and breadth, and is called the horizontal projection or *plan*.

The other of these drawings is always taken vertically, parallel to the height, and is called the vertical projection or *elevation*.

As it seldom happens that the object to be described is of such dimensions that it can be conveniently drawn in its actual size, it becomes necessary to first understand what is meant by *Scale Drawing*.

CHAPTER I.—Scale Drawing.

Scale Drawings are of three kinds: Full Size, Reduced, and Enlarged.

In a **Full Size** scale drawing, every line in the object is described at its full length.

In a **Reduced** scale drawing, every line in the

object is described a certain, constant number of times smaller than it is.

In an **Enlarged** scale drawing, every line in the object is described a certain, constant number of times larger than it is.

Section I.—Full Size Drawing.

Prob. 1.—Represent a horizontal line 5 in. long. Full size.

Prob. 2.—Represent two parallel vertical lines $3\frac{5}{8}$ in. and $2\frac{7}{8}$ in. in length and $\frac{1}{2}$ in. apart. Full size.

Prob. 3.—Draw a circle $1\frac{1}{8}$ in. diameter. Full size.

Prob. 4.—Draw a $2\frac{1}{4}$ in. square. Full size.

Prob. 5.—Draw a $1\frac{3}{4}$ in. square, diagonal vertical. Full size.

Prob. 6.—Draw a parallelogram 2 in.$\times \frac{3}{4}$ in., longer sides 30° with horizon to the right. Full size.

Section II.—Reduced Scale Drawing.

Prob. 1.—Represent a line 8 ft. long. Reduced; scale, $\frac{1}{4}$ in. to the foot.

SOLUTION.—The statement, "Scale $\frac{1}{4}$ in. to the foot," signifies that 1 ft. is the unit of measure, and $\frac{1}{4}$ in.

A———————————————B

Scale, ¼ in. to the ft.

is the unit of representation. As there are 8 units of measure there must be 8 units of representation. Hence, *AB* represents 8 ft. because it is $\frac{3}{4}$ in. in

length (eight units of representation), and the scale
is ¼ in. to the foot.

Prob. 2.—Draw a line that will represent 25 ft.
Reduced; scale, ¼ in. to the foot.

Prob. 3.—Represent 100 ft. in length. Reduced;
scale, $\frac{1}{16}$ in. to the foot.

Prob. 4.—Represent 5 ft. in length. Reduced;
scale, 2 ft. to the in.

Prob. 5.—Represent one side of the school-room
door. Reduced; scale, ½ in. to the foot.

Rule.—Multiply the number of units of measure
in the line to be represented by the unit of representation, and the product will be the number of
units of representation in the required line.

Section III.—*Enlarged Scale Drawing.*

Prob. 1.—Represent ⅜ in. in length. Enlarged;
scale, ⅔ in. to ⅛ in.

SOLUTION.—In this problem, the unit of measure
is ⅛ in. and the
unit of representa-
tion is ⅔ in. Hence,

A————————————————————B
Scale, ⅔ in. to ⅛ in.

to represent ⅜ in. in length would require a line
$3 \times$ ⅔ in. $=$ ⁶⁄₃ $= 2$ in. *AB*, which is 2 in. long, represents ⅜ in. when the scale is ⅔ in. to the ⅛ in.

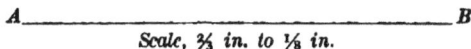

Prob. 2.—Represent ¾ in. in length. Enlarged;
scale, ⅞ in. to the ¼ in.

Prob. 3.—Represent ⅜ in. in length. Enlarged; scale, ½ in. to ⅛ in.

Prob. 4.—Represent 1/16 in. in length. Enlarged; scale, 32 in. to 1 in.

Prob. 5.—Represent 23/100 in. in length. Enlarged; scale, ⅛ in. to 1/100 in.

Prob. 6.—Represent a circle 3/25 in. in diameter. Enlarged; scale, ⅜ in. to 1/25 in.

Prob. 7.—Represent an equilateral triangle; side, 9/50 in. in length. Enlarged; scale, ⅛ in. to 1/50 in.

Prob. 8.—Represent a regular octagon 3/25 in. in diameter. Enlarged; scale, 1 in. to 1/25 in.

Rule.—Multiply the number of units of measure by the unit of representation, and the product will be the length of the required line in the drawing.

CHAPTER II.—Plans.

A **Plan** is a drawing which describes the length and breadth of an object on a surface that is considered as always lying horizontal.

Plans are of two kinds: parallel plans and angular plans.

A **Parallel** plan is one that describes only lines and surfaces that are parallel to itself.

An **Angular** plan is one in which the horizontal dimensions of lines or surfaces that are not parallel to itself are described.

Section I.—Parallel Plans.

Prob. 1.—Draw a plan of a rectangular block that is 10 ft. long and 8 ft. wide; scale, $\frac{1}{16}$ in. to the foot.

SOLUTION.—The block is rectangular; therefore, the plan will be rectangular. It is 10 ft. long and 8 ft. wide, and is to be represented on a scale $\frac{1}{16}$ in. to the foot; therefore, the plan will be a rectangle $\frac{10}{16}$ in. long and $\frac{8}{16}$ in. wide. Draw a rectangle, *ABCD*, (Fig. 40), $\frac{5}{8}$ in. long and $\frac{1}{2}$ in. wide, and it will be the required plan of the block.

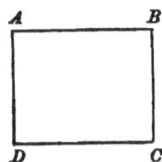

Fig. 40.

Prob. 2.—Draw the plan of a rectangular block 13 ft. long and 11 ft. wide; scale, $\frac{1}{4}$ in. to the foot.

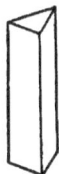

Prob. 3.—Draw the plan of a triangular prism standing on end (Fig. 41), the sides of which are 2 ft., 2$\frac{1}{2}$ ft., and 3 ft. wide; scale, $\frac{3}{8}$ in. to the foot.

Fig. 41.

Prob. 4.—Draw the plan of a circular cone. Base of cone, 19 ft.; scale, $\frac{3}{16}$ in. to the foot. (Fig. 42).

Prob. 5.—Draw the plan of a rectangular block $\frac{1}{16}$ in. wide and $\frac{1}{4}$ in. long; scale, 1 in. to $\frac{1}{4}$ in.

Fig. 42.

Prob. 6.—Draw a plan of a sphere (ball) 8000 miles in diameter; scale, $\frac{1}{8}$ in. to the 1000 miles.

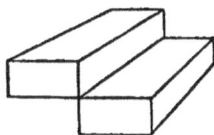

Prob. 7.—Draw the plan of two rectangular blocks, 3 ft. long and 1 ft. wide, arranged as shown in the cut (Fig. 43). Scale, $\frac{1}{2}$ in. to the foot.

Fig. 43.

Prob. 8.—Draw the plan of a flight of six steps, each step 4 ft. long and 10½ in. wide; scale, ½ in. to the foot.

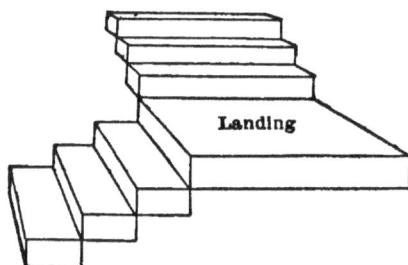

Fig. 44.

Prob. 9.—Draw the plan of a flight of 6 steps, with square landing, making a quarter turn midway of the flight. Steps, 3 ft. × 10½ in.; landing, 3 ft. square; scale, ¼ in. to the foot. (Fig. 44).

Prob. 10.—Draw the plan of a flight of 9 steps, with two square landings, making two quarter turns, 3 steps between the turns. Steps, 4 ft. × 1 ft.; landings, 4 ft. square; scale, ½ in. to the foot.

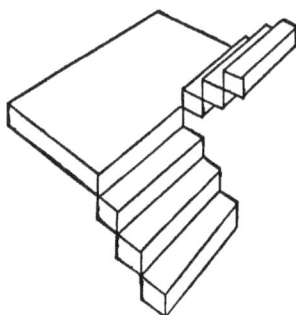

Fig. 45.

Prob. 11.—Draw a plan of 6 steps (Fig. 45), making a half turn in the middle by a landing 4 ft. × 8 ft. Steps, 1 ft. × 4 ft.; scale, ½ in. to the foot.

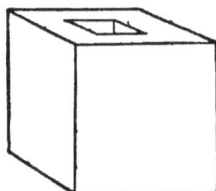

Fig. 46.

Prob. 12.—Draw the plan of a cubical block (Fig. 46), 4 ft. square, with a hole 18 in. square, extending through its center from top to bottom. Scale, ¼ in. to the foot.

Def.—A cube is a solid having six square faces.

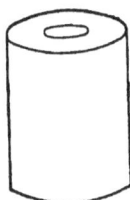

Fig. 47.

Prob. 13.—Draw the plan of a cylinder (Fig. 47), 24 in. diameter, 16 in. long, with a round hole 13 in. in diameter, extending lengthwise through its center; scale, $\frac{1}{16}$ in. to the inch.

Prob. 14.—A grindstone lying on the ground is 5 ft. in diameter, with a hole 6 in. square in the center. Draw the plan; scale, $\frac{1}{2}$ in. to the foot.

Prob. 15.—Draw the plan of a cylinder lying on its side. Diameter, $\frac{1}{32}$ in.; length, $\frac{5}{16}$ in.; scale, 16 in. to the inch.

Prob. 16.—Draw the plan of a flight of stairs consisting of 2 flyers and 3 winders, making a quarter turn. Treads of flyers, 4 ft. × 1 ft.; radius of inner curve of winders, 1 ft.; radius of outer curve of of winders, 5 ft.; scale, $\frac{1}{2}$ in. to the ft. (Fig. 48).

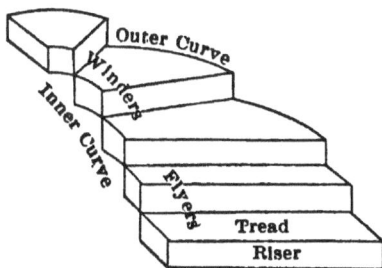

Fig. 48.

Prob. 17.—Draw the plan of a flight of 6 steps, 4 flyers and 2 winders, making a quarter turn. Treads of flyers, 3 ft. × 10½ in.; outer radius of winders, 5 ft. 6 in.; inner radius, 18 in.; scale, $\frac{1}{4}$ in. to the foot.

Prob. 18.—Draw the plan of a flight of 6 steps (Fig. 49), 3 flyers and 3 winders, making a quarter turn to the left hand. Treads of flyers, 10½ in. × 3 ft. 6 in. Radius of outer circle of winders, 4 ft. 6 in.; scale, $\frac{1}{4}$ in. to the foot.

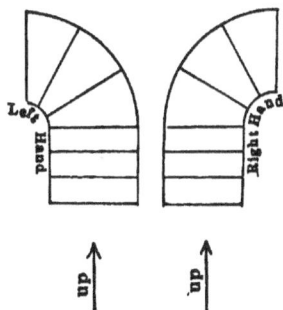

Fig. 49.

Prob. 19.—Draw the plan of a flight of 10 steps, 4 winders, making a half turn to the right hand, and 3 flyers at either end of the flight. Treads of flyers, 3 ft. × 1 ft. Radius of inner curve of winders, 9 in.; scale, ¼ in. to the foot.

Prob. 20.—Draw the plan of a flight of 12 steps, making two quarter turns to the left-hand; 3 winders in each turn, with 2 flyers between them, and 2 flyers at either end of the flight. Treads of flyers, 3 ft. × 10½ in. Radius of outer curve of winders, 4 ft.; scale, ⅜ in. to the foot.

Prob. 21.—Draw the plan of a flight of 12 steps, all winders, making a whole turn. Length of tread, 2 ft., 10½ in.; outer radius of winders, 4 ft.; scale, ½ in. to the foot.

Section II.—Plans of the same object at different levels.

When the horizontal dimensions of any object are the same at all heights, then the length and breadth of the object may be described *in one plan.*

When the horizontal dimensions of any object vary at different heights, as in houses of more than one story, ships of more than one deck, etc., then the length and breadth of the object can be described only by *as many different plans as there are different levels at which these dimensions change.*

The different plans required to describe all the horizontal dimensions of any object have each a name appropriate to the part of the object described.

In buildings, we may have a ground plan, a basement plan, a story plan, and a roof plan.

A **Ground Plan** describes the length and breadth of the ground covered.

A **Basement Plan** describes the length and breadth of the cellar.

A **Story Plan** shows the length and breadth of every part of a building at a given story.

A **Roof Plan** describes the length and breadth of every part of the roof of the object.

Prob. 1.—Draw the ground plan of the school-house; scale, $\frac{1}{8}$ in. to the foot.

Prob. 2.—Draw the plan of the school-house grounds, and the ground plans of the buildings on it; scale, 64 ft. to the inch.

Prob. 3.—Draw the ground plan of your school desk; scale, 1 in. to the foot.

Note.—The ground plan of a portable object is the plan of that part that touches the ground or floor.

Prob. 4.—Draw the ground plan of your teacher's desk; scale, $\frac{1}{2}$ in. to the foot.

Prob. 5.—Draw the ground plan and seat plan of a common wood-seat chair; scale, $1\frac{1}{2}$ in. to the foot.

Prob. 6.—Draw the first-floor plan of your school-house; scale, $\frac{1}{16}$ in. to the foot.

Prob. 7.—Draw the ground plan and the roof plan of two towers (Fig. 50), 34 ft. outside diameters, 48 ft. high, 38 ft. between centers, and connected by a covered passage-way 12 ft. wide and 12 ft. high, outside measurement; walls of towers, 24 in. thick, and walls of passage, 12 in. thick; scale, $\frac{1}{16}$ in. to the foot.

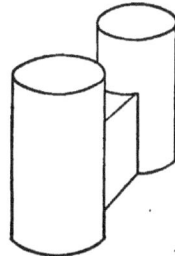

Fig. 50.

Prob. 8.—Draw the ground plan of the doorway of your school-room, with the door half open; scale, ½ in. to the foot.

Prob. 9.—Draw the ground plan of a house (Fig. 51), 14 ft. × 30 ft., with 10 ft. projection, 12 ft. long, in the middle of longer side; walls, 9 in. thick; scale, ¼ in. to the foot.

Fig. 51.

Prob. 10.—Draw the ground plans of two buildings (Fig. 52), 48 ft. × 60 ft., with a hexagonal tower in the middle of one of the longer sides, and projecting one half beyond the wall. One of these buildings is of brick, with a foundation wall 24 in. thick; and one is a frame building, with a foundation wall 16 in. thick; scale, ⅛ in. to the foot.

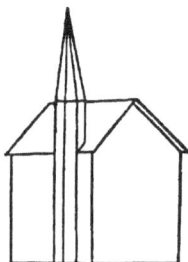

Fig. 52.

Prob. 11.—A house 16 ft. × 20 ft. has a rectangular projection at one corner (Fig. 53). This projection has 2 walls, 5 ft. apart and 4 ft. long, making angles of 135° with the end and side walls of the house. Draw the ground plan; walls, 15 in. thick; scale, ¼ in. to the foot.

Fig. 53.

Prob. 12.—A house is 16 ft. × 36 ft., and has a bay window in the middle of one of the longer sides. This window, which has three equal sides, is 10 ft. wide and projects 3 ft. (Fig. 54.) All the walls are 1 ft. thick. Draw a ground plan; scale, ¼ in. to the foot.

Fig. 54.

Prob. 13.—Draw the ground plan of a house 18 ft. × 28 ft., extreme dimensions. One end is divided into 3 equal spaces, making 120° angles with each other and with the side walls of the house; scale, ¼ in. to the foot. (Fig. 55.)

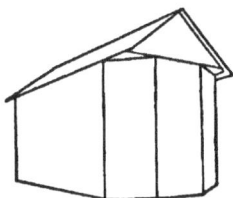

Fig. 55.

Prob. 14.—A house 24 ft. square has a projection 12 ft. square at one end of one side, and in the angle formed by the house and the projection is a porch 12 ft. sq. (Fig. 56). Walls of house, 15 in. thick, and the outermost corner of the porch is supported by a pier 12 in. square. Draw ground plan of house and porch; scale, ¼ in. to the foot.

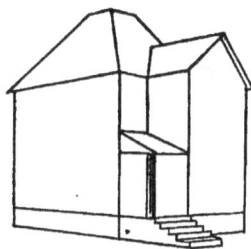

Fig. 56.

Prob. 15.—Draw the ground plan of your own house, or of any house that you may design.

Prob. 16.—Draw the ground plan of an arch, 12 ft. span, standing on two piers, 3 ft. square; scale, ¼ in. to the foot.

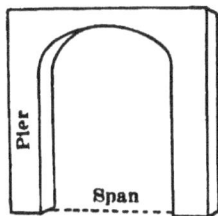

Fig. 57.

Section III.—Angular Plans.

Prob. 1.—Draw the plan of a gable roof 21 ft. long and 19 ft. span; scale, $\frac{1}{32}$ in. to the foot.

SOLUTION.—The length of a gable roof is measured on the eaves, and, unless otherwise specified, the eaves are horizontal, opposite, parallel, and equal, and the span is the perpendicular distance between

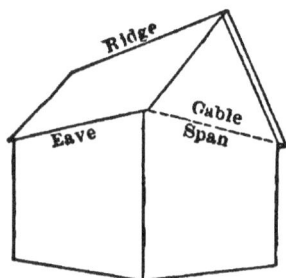

Fig. 58.

them (Fig. 58). As the roof described in this problem is 21 ft. long and 19 ft. span, and as the scale is $\frac{1}{32}$ in. to the foot, we draw the rectangle $ABCD$ (Fig. 59), $\frac{21}{32}$ in. $\times \frac{19}{32}$ in., and as the ridge is equally distant from the eaves, we draw EF parallel to the lines representing the eaves and midway between them. $ABFCDE$ is the required plan of the roof, because it exactly describes the length and breadth of every part of the roof *measured horizontally.*

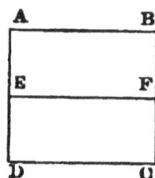

Fig. 59.

Prob. 2.—Draw the plan of a gable roof 18 ft. long and 14 ft. span; scale, $\frac{1}{8}$ in. to the foot.

Fig. 60.

Prob. 3.—A gable-roofed house (Fig. 60), 30 ft. long and 18 ft. wide, has a wing in the middle of one of its sides, 16 ft. long and 6 ft. projection, covered by an extension of the roof of the main house; roof projection beyond the wall of the house on all sides, 1 ft. Draw the plan of the roof; scale, $\frac{1}{4}$ in. to the foot.

Prob. 4.—Draw the plan of a rectangular pyramid with the apex vertically over one corner of the base (Fig. 61). Base, 9 ft. $\times$ 6 ft. sq.; scale, $\frac{1}{4}$ in. to the ft.

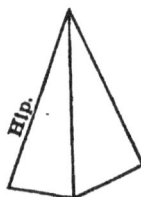

Fig. 61.

Prob. 5.—Draw the plan of a frustum of a rect-angular pyramid. Base, 11 ft. ×
13 ft.; and deck, 6 ft. × 6 ft. 6 in.
Hips, equal in length; scale, ⅛ in.
to the foot. (Fig. 62.)

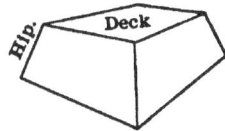

Prob. 6.—Draw the plan of a
hooded-gable roof. Ridge, 19 ft.;
eaves, 30 ft. Span of gable at the
top, 8 ft. Span at the bottom, 18
ft. Scale, $\frac{1}{16}$ in. to the foot. (Fig.
63.)

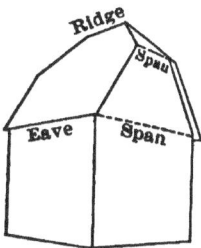

Fig. 62.

Fig. 63.

Prob. 7.—Draw a plan of the roof
of a gable-roofed house with wing.
House, 20 ft. × 24 ft. Wing, 8 ft.
projection, 16 ft. in length,
and in the middle of the 20
ft. side of the house. Eaves
and the gable of the roof
project 18 in.; scale, $\frac{1}{16}$ in. to
the foot. (Fig. 64.)

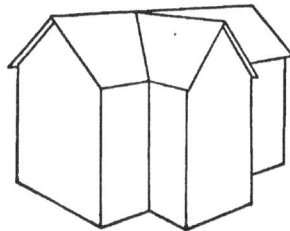

Prob. 8.—Draw the plan of
the roof of a gable-roofed

Fig. 64.

house and wing. The ridge of
the roof of the wing, even with
the eaves of the main house.
Main roof, 16 ft. span and 24 ft.
long. Wing roof, 12 ft. square,
and in the middle of the longer
side of the house; scale, $\frac{1}{16}$ in.
to the foot. (Fig. 65.)

Fig. 65.

Def.—A rafter is a roof timber extending from
the eaves to the ridge.

Prob. 9.—Draw the plan of a gable-roofed house with curved rafters. Eaves, 16 ft. long; span, 14 ft.; scale, $\frac{1}{16}$ in. to the foot. (Fig. 66.)

Prob. 10.—Draw the roof plan of a gable-roofed building with smaller gable-roofed extension at one end. Main house, 20 ft. wide and 24 ft. long. Extension, 12 ft. square, 6 ft. from right-hand side of end. Roof projection, 18 in. on all sides; scale, $\frac{1}{8}$ in. to the ft. (Fig. 67.)

Fig. 66.

Prob. 11.—Draw the top plan of your school desk; scale, 1 in. to the foot.

Fig. 67.

Section IV.—Condensed Plans.

Whenever it can be done without rendering the description any less clear and unmistakable, it is the custom among draughtsmen to describe all the horizontal dimensions belonging to the same object, in one drawing, called a *condensed plan.*

Of the different plans which constitute a condensed plan, one is called the Principal Plan.

The **Principal Plan** may be the plan of the object at any level, and is always drawn in a continuous line.

Plans of levels above the principal plan may be drawn in lines made up of dashes and hyphens; as, — · — · — · or — · · · — · · · or — · · · — · · , etc.

Plans of levels below the principal plan may be drawn in lines made up of dashes and dots; as, — · — · — · or — · · · — · · · or — · · · — · · · , etc.

Prob. 1.—Draw a condensed ground plan of a box 2 ft. 9 in. sq., made of planks 3 in. thick, and having a lid projecting 6 in. all around; scale, ¼ in.

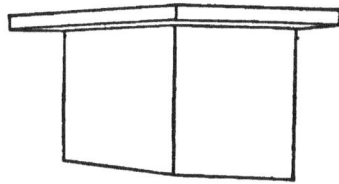

Fig. 68.

to the foot. (Fig. 68.)

SOLUTION.—In a condensed ground plan, the ground plan of the box is a continuous line, and the plans of the lid and inside are drawn in dotted lines, *A*, Fig. 69.

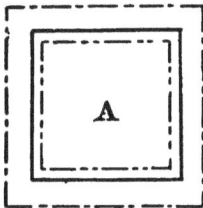

Scale ¼ in.to the ft.

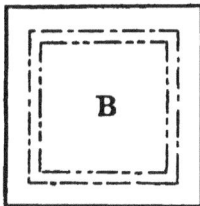

Fig. 69.

If the problem called for a condensed lid plan, it would be drawn as shown in *B*, Fig. 69.

Prob. 2.—Draw a condensed middle plan of a common spool; scale, 2 in. to the in. (Fig. 70.)

Prob. 3.—Draw a condensed lid plan of a box 4 ft. wide and 6 ft. 3 in. long, made of 3 in. plank, with lid projecting 6 in. all

Fig. 70.

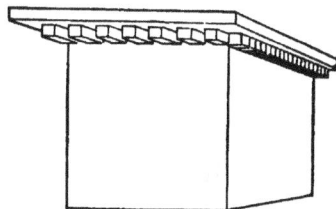

Fig. 71.

round, and having a row of rectangular blocks (dentils) 3 in. square and 6 in. apart on all sides immediately below the lid; scale, ¼ in. to the foot. (Fig. 71.)

Prob. 4.—Draw a condensed ground plan of a cubical block 16 ft. square, with a hole 9 ft. square extending through its center from top to bottom, and an opening 3 ft. wide and 6 ft. high extending through one of its sides at the middle of the lower edge; scale, $\frac{1}{16}$ in. to the foot. (Fig. 72.)

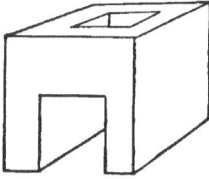

Fig. 72.

Prob. 5.—Draw a condensed ground plan of a frame-work 4 ft. wide, 7 ft. long, and 3 ft. high, made of timbers 15 in. square; scale, $\frac{1}{4}$ in. to the foot. (Fig. 73.)

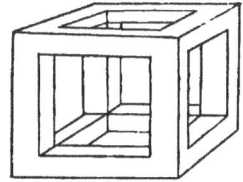

Prob. 6.—Draw a condensed middle plan of a block 10 ft.

Fig. 73.

square and 5 ft. high, with panels on each of the four vertical sides, 5 ft. long, 2 ft. 6 in. high, and 6 in. deep, beveled on the edge of the panel opening at an angle of 45°; scale, $\frac{1}{8}$ in. to the foot. (Fig. 74.)

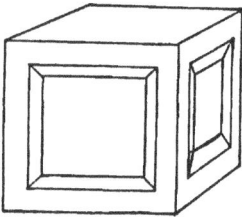

Fig. 74.   **Def.**—A bevel is a slanting edge.

Prob. 7.—Draw a condensed ground plan of a bobbin. Shank, $1\frac{1}{4}$ in. in diameter, $6\frac{3}{4}$ in. long between the flanges, and a hole $\frac{5}{8}$ in. diameter lengthwise through the center; ends, $4\frac{3}{4}$ in. diameter, $\frac{3}{4}$ in. thick, and square edge; scale, $\frac{1}{2}$ in. to the inch. (Fig. 75.)

Prob. 8.—Draw a floor plan of the school-room; scale, $\frac{1}{4}$ in. to the foot.

Fig. 75.

NOTE.—There are cases in actual practice in which certain facts of construction are represented in a way peculiar to themselves, without regard to any rule of constructive drawing. This is true of windows; for, in a strictly · floor plan, the windows would not be represented at all, because they do not come to the floor. For convenience sake, draughtsmen have adopted the plan of representing all windows as shown in *W*, and all doors as shown in *D*, Fig. 76.

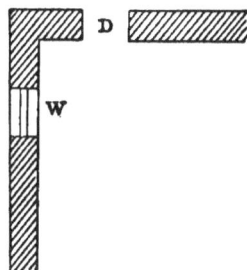

Fig. 76.

Fig. 77.

Prob. 9.—Draw the plan of a wall 3 ft. thick, 25 ft. long and 15 ft. high, pierced by two equal openings 8 ft. wide, having a semicircular arch at the top, and 8 ft. high to the springing of the arch; scale, ¼ in. to the foot. (Figs. 77 and 78.)

Fig. 78.

NOTE.—Whenever it is not expressly stated, it is expected that the draughtsman will exercise his judgment in condensing his plans. His whole aim should constantly be to state his facts as briefly, compactly, and clearly as possible, without regard to time, labor, or the number of drawings required.

CHAPTER III.—ELEVATIONS.

An **Elevation** is a drawing made on a flat surface that is regarded as always standing vertical—perpendicular to the plan.

Heights or **Vertical Dimensions** are always described by the elevation because the elevation itself is always vertical.

The lines and surfaces that may be described by an elevation are not always vertical, and therefore they are not always parallel to it. Hence, there are two kinds of elevations: parallel elevations and angular elevations.

A **Parallel Elevation** describes only lines and surfaces that are parallel to itself.

An **Angular Elevation** describes lines and surfaces that are not parallel to itself.

Both parallel and angular elevations may describe interior or exterior heights, and it not infrequently happens that to completely describe the heights of all parts of an object, many different elevations of both the exterior and interior are required.

Each elevation must have a title inscribed upon its face indicating the side and part of the object described.

To describe stationary objects, the elevations may be named according to the points of the compass, or indicated as front, rear, end, or side elevations.

Elevations describing portable objects must have appropriate names devised by the draughtsman.

Section I.—Exterior Parallel Elevations.

Fig. 79.

Prob. 1.—A house, 13 ft. × 16 ft., with walls 6 in. thick and 11 ft. high, including foundation 1 ft. 6 in., and with a gable roof, 6 ft. pitch (vertical height), projecting 9 in. beyond the walls on all sides, stands with its longer sides north and south. There is a door, 2 ft. 6 in. × 6 ft. 6 in., 2 ft. from

the westerly end in the south wall. The windows
are all 2 ft. 3 in. × 5 ft., 2 ft. 3 in. above the top of
the foundation. In the south wall is one window
midway between the door and the easterly end. In
the east wall are two windows, 3 ft. apart, and
equally distant from the ends. In the north wall
there is one window, in the middle. In the west
wall are two windows, 2 ft. 3 in. from the ends.
Draw the plan and elevations of each side; scale,
$\frac{1}{32}$ in. to the foot. (Fig. 79.)

SOLUTION.—Regarding the top of the drawing as
north, draw a condensed
plan, A, (Fig. 80), $\frac{13}{32}$ in.
wide by $\frac{16}{32}$ in. long, the
longer side extending
north and south, and lo-
cate in it the width and
position of the door and
windows. From each side
of the plan draw perpen-
dicular leading (dotted)
lines. Perpendicular to
each set of leading lines,

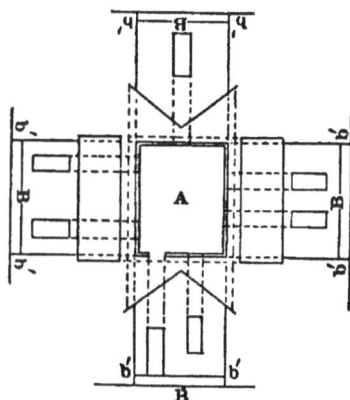

Scale, $\frac{1}{32}$ in. to the foot.

Fig. 80.

and at a distance from the plan greater than the
entire height of the object to be represented, draw
the base-line $b'b'$. From this base-line lay off
towards the plan on each leading line the height of
the object at that point, and at these points draw
the necessary connecting lines to complete the
several elevations, B, B, B, and B.

Prob. 2.—Draw the plan and parallel elevations
of a cubical block 4 ft. square; scale, $\frac{1}{4}$ in. to the ft.

Prob. 3.—Draw the plan and parallel elevations of the end and side of a rectangular block 6 ft. 6 in. long, 5 ft. wide, and 4 ft. high; scale, ½ in. to the ft.

Prob. 4.—Draw the plan and parallel end and side exterior elevations of a rectangular block 3 ft. 7½ in. long, 4 ft. 4½ in. wide, and 4 ft. high, with two holes 19½ in. square, one 3 ft. 7½ in. long, and the other 4 ft. long, perpendicular to each other and to the faces of the block, and passing through its center; scale, ½ in. to the foot. (Fig. 81.)

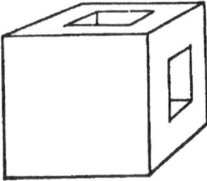

Fig. 81.

Prob. 5.—Draw the plan and the parallel elevation of two adjacent sides of two timbers, laid at right angles each across the middle of the other. The timbers are each 18 in. wide and 6 in. thick; one of them is 6 ft. long, and the other is 12 ft. long; scale, ¼ in. to the foot. (Fig. 82.)

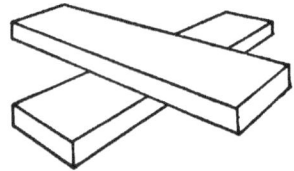

Fig. 82.

Prob. 6.—Draw the plan and the parallel end and side elevations of a box 13½ ft. long, 5½ ft. wide, and 6 ft. high, with a top 3 in. thick, projecting 6½ in., with 2 dentils 3 in. square at each corner, immediately below the top; scale, ¼ in. to the foot. (Fig. 83.)

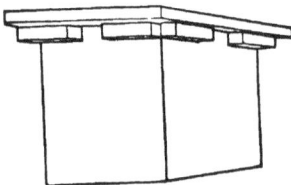

Fig. 83.

Prob. 7.—Draw the plan and the parallel end and side elevations of the teacher's platform; scale, ¼ in. to the foot.

Prob. 8.—Draw plan and parallel end and rear elevations of your school desk; scale, 1 in. to the foot.

Prob. 9.—Draw the plan and the end and front elevations of two steps. Treads, 1 ft. × 3 ft., and risers, 6 in.; scale, ½ in. to the foot. (Fig. 84.)

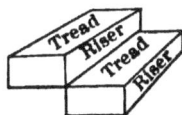

Fig. 84.

Prob. 10.—Draw the plan and the elevation of a flight of 6 steps, making a quarter turn in the middle, to the right hand, by a square landing. Treads, 1 ft. × 4 ft.; risers, 7½ in.; landing, 4 ft. square; scale, ½ in. to the foot.

Prob. 11.—Draw the plan and the parallel end and side elevations of a wall 3 ft. thick, 15 ft. high, and 25 ft. long, pierced by two circular arched openings, 8 ft. wide, and 8 ft. high to the springing of the arches; scale, ¼ in. to the foot.

Prob. 12.—Draw the plan and the parallel front and side exterior elevations of a flat-roofed house 16 ft. × 24 ft.; height of wall, 14 ft., including the foundation of 2 ft.; roof-cornice extending across the front, 1 ft. projection, supported by 9 brackets, 10½ × 12 in., and equally distant apart; the chimney, 24 in.× 18 in., and extending 4½ ft. above the roof, is in the rear wall, 1 ft. from the right-hand side of window; there is a window in the middle of both the longer sides, and one in the middle of one end; a door is in the middle of the front. The windows are 2 ft. 6 in. × 6 ft., and 2 ft. 6 in. from the floor; the door is 3 ft. wide and 7 ft. high; scale, ¼ in. to the foot.

Prob. 13.—Assume the dimensions and draw the plan and the elevation of a gable-roofed house with front and side porches.

Section II.—*Interior Parallel Elevations.*

Prob. 1.—A school-room is 32 ft. wide, 36 ft. long, and 12 ft. high. In the front wall there are two doors 3 ft. wide and 7 ft. high, 2 ft. 6 in. from the end of the wall; in each of the side walls are 3 windows, 3 ft. wide × 6 ft. 6 in. high, 3½ ft. from the floor, and 6 ft. 6 in. apart. There is a blackboard 4 ft. wide and 2 ft. 6 in. from the floor on the front wall between the doors, and another extending entirely across the rear wall. Draw the plan and the parallel interior elevation of each wall; scale, $\frac{1}{64}$ in. to the foot.

SOLUTION.—Draw the plan, $\frac{32}{64}$ in. wide and $\frac{36}{64}$ in.

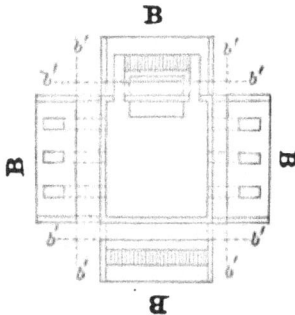

Fig. 85.

long, showing the widths and positions of the doors and windows, *A* (Fig. 85). Draw from each side of this plan leading lines, and perpendicular to these draw the base-lines *b'b'*. From these lines lay off on the leading lines, away from the plan (see note), the heights of the different parts of the wall above the floor. Draw lines joining these points, and they will enclose the elevations of the several walls *B*, *B*, *B*, and *B*.

NOTE.—Compare this solution with that on page 63, and it will be seen that the only difference between an interior and an exterior elevation is that in an interior elevation the base-line is the line of the elevation that is nearest to the plan, and in an exterior elevation the base-line is the line of the elevation that is farthest from the plan.

Prob. 2.—Draw the plan and the interior elevation of the front wall of the school-room and the teacher's platform; scale, ⅛ in. to the foot.

Prob. 3.—Draw the plan and the interior elevation of the rear wall of the school-room and the pupils' desks; scale, ¼ in. to the foot.

Prob. 4.—A room, 25 ft. × 21 ft., and 10 ft. high, has five windows, 30 in. wide, and 6 ft. 6 in. high, 2 ft. from the floor; two doors, 36 in. wide and 7 ft. 6 in. high; and a chimney, 2 ft. breast (width across the front) and 1 ft. jamb (projection from the wall). The length of the room stands north and south. The chimney is in the middle of the east wall, with a window on either side and 3 ft. from it; one door is opposite the chimney, and the other door is in the middle of the south wall, with a window on either side, and 2 ft. 6 in. from it; and one window is in the middle of the north wall. Draw the plan and the interior elevation of each wall; scale, ⅛ in. to the foot.

Prob. 5.—Draw the floor plan and the interior parallel elevations of each side of your school desk; scale, 1 in. to the foot.

Prob. 6.—Draw the plan and four interior parallel elevations of the walls of any hall in the school building; scale, ¼ in. to the foot.

Section III.—Condensed Elevations.

As in making plans time and space may be economized, when the object represented is not too complicated, by drawing two or more plans within the same outline, so in making elevations when

there are not too many parts to be represented, exterior and interior elevations may be made in the
same drawing.

Prob. 1.—Draw a condensed plan and elevation
of a box 8 in. square and 7 in. high,
outside measurement, and made of boards
1 in. thick. The sides are mitered, the
bottom is let in, and the top is laid on;
scale, ⅛ in. to the inch.

Fig. 86.

SOLUTION.—Draw the plan *ABCD*, ⅝ in. square =
1 in. As the sides extend to
the ground, the plan of the inside of the box will be drawn
in full lines, *EFGH*, and ⅛ in.
from the sides of *ABCD*, because the sides are 1 in. thick.
The joints between the sides at
the corners will be represented
by diagonal lines, *AI*, *FB*, *GC*,
and *HD*, joining the plan of
the outside and inside of the
box. Draw the elevation *dcih*,
⅞ in. high. Draw a full line
jk, ⅛ in. below the top, to represent the bottom of the lid.

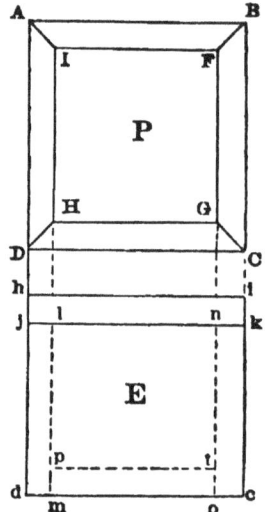

Fig. 87.

Draw *lm* and *no* to represent the interior elevation
of the side, and *pt* to represent the interior elevation of the bottom, showing that it is let in. *E* is
the elevation and *P* is the plan, and together they
completely and clearly describe the dimensions and
construction of the box required in the problem.
(Fig. 87.)

Prob. 2.—Draw the plan and as many elevations as may be necessary to completely and clearly describe a box 31 in. square and 17 in. high, made of planks 3 in. thick, and having no side let in at the ends; scale, ⅛ in. to the inch. (Fig. 88.)

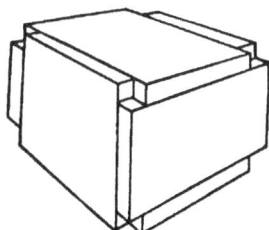

Prob. 3.—A brick wall 16 in.

Fig. 88.

thick and 12 ft. high encloses a space 24 ft. × 30 ft. The shorter walls are blank; there is a door 3 ft. × 7 ft. 6 in., 1 ft. from the ground, reached by two steps, 3 ft. × 1 ft. tread, and 6 in. rise; and there are two windows 2 ft. 10½ in. wide, 6 ft. 6 in. long, and 2 ft. from the floor, dividing the opposite wall into three equal spaces. Make one plan and elevation which completely describes the construction; scale, ¼ in. to the foot.

Prob. 4.—Draw one plan and one elevation that completely describes the dimensions and construction of a tool box or a work box; scale, ¼ in. to the foot.

Section IV.—Angular Elevations.

An **Angular Elevation** describes lines and surfaces that are not parallel to itself.

Prob. 1.—Draw an angular elevation of two sides of a cube 4 ft. square; scale, ⅛ in. to the foot.

SOLUTION.—Draw a ½ in. square to represent the plan of a cube 4 ft. square. Draw *AB*, the base-line of the elevation, at an angle with the sides

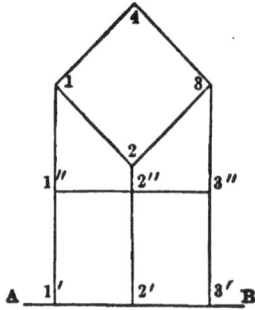

Fig. 89.

1-2 and 2-3. From 1, 2, and 3 draw leading lines perpendicular to the base-line *AB*. On these lines lay off, one inch from the base-line, 1'1", 2'2", and 3'3" for the height. Draw 1"2"3", and 1'1"2"3"3'2' is the outline of an angular elevation of two sides of a cube 4 ft. square. (Fig. 89.)

Prob. 2.—Draw an end, a side, and an end and side elevation of a house 16 ft. wide, 23 ft. long, and 14 ft. high, including the foundation, which is 2 ft. high. Gable roof, 8 ft. pitch and 1½ ft. projection; bay window in middle of end, 10 ft. wide, 11 ft. high, including foundation, 4 ft. projection, 3 equal sides, and gable roof ½ pitch; scale, $\frac{1}{32}$ in. to the foot.

SOLUTION.— Draw the ground plan, *abcd*, $\frac{16}{32}$ in. wide and $\frac{23}{32}$

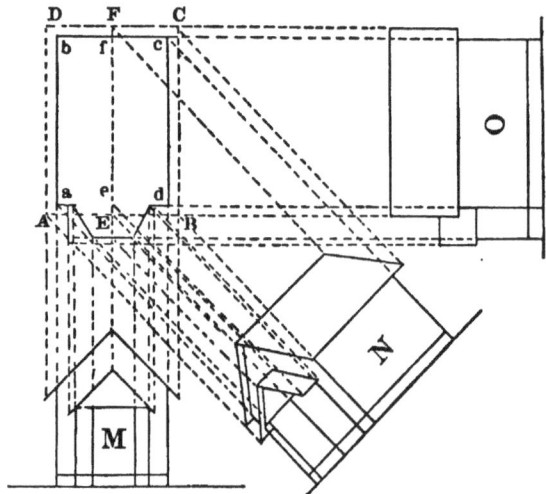

Fig. 90.

in. long, showing the angle and projection of the bay window, *E*. Draw the roof plan, *ABCD*, $\frac{19}{32}$

in. wide and $\frac{24}{32}$ in. long in dotted line. Draw leading lines from the plan, and construct the elevations *M, N, O.* (Fig. 90.)

Prob. 3.—Draw a plan and elevation of a pyramid
4 ft. square at the base and 6 ft. pitch; scale, $\frac{1}{4}$ in.
to the foot.

Prob. 4.—Draw a plan and elevation of a hexagonal prism 18 in. in diameter and 14 in. long;
scale, $\frac{1}{16}$ in. to the inch.

Prob. 5.—Draw a plan and elevation of a cone 3
in. diameter and 1½ in. pitch; scale, 1 in. to the
inch.

Prob. 6.—Draw a plan and elevations of a sphere
$\frac{1}{32}$ in. in diameter; scale, 1 in. to $\frac{1}{32}$ in.

NOTE.—The preceding problems require the description of
objects so formed that a parallel elevation can not be made,
and therefore the term angular elevation is omitted.

Prob. 7.—Draw a plan and angular elevation of a
cube 5 ft. 6 in. square; scale, $\frac{1}{4}$ in. to the foot.

Prob. 8.—Draw a plan and angular front and side
elevation of your school desk; scale, 1 in. to the
foot.

Prob. 9.—Draw a plan and a front elevation, a
side elevation, and a front and side elevation of a
house 21 ft. wide and 33 ft. long; 14 ft. high, including a foundation of 18 in.; flat roof; bay window, 10 ft. wide, 3 ft. 6 in. projection, 3 equal sides;
scale, $\frac{1}{8}$ in. to the foot.

Prob. 10.—Draw a plan and an interior angular
elevation of a corner of the school-room, taking in
12 ft. of the walls forming the corner; scale, $\frac{1}{8}$ in.
to the foot.

Prob. 11.—A room is 21 ft. long, 15 ft. wide, and 10 ft. high; it has seven windows 3 ft. wide, 6 ft. high, and 3 ft. from the floor; two doors 3 ft. wide, 7 ft. high; one chimney 5 ft. wide (breast), 18 in. projection (jamb),—chimney in the middle of the shorter side; three windows 3 ft. apart and 3 ft. from the end, in the longer wall to the right hand of chimney; one window opposite chimney, and one door on either side of it and 3 ft. from it; one window in the middle of the longer wall to the left hand of chimney. Draw plan and angular elevations of each corner, taking in the entire walls forming each ; scale, ⅛ in. to the foot.

Prob. 12.—Draw a plan and angular elevation of a chimney 6 ft. breast and 21 in. jamb, with mantle 10½ in. wide, and 3 in. thick, 5 ft. from floor, and fire-place opening 2 ft. square and 9 in. jambs; scale, ½ in. to the foot.

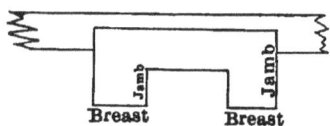
Fig. 91.

Prob. 13.—Draw a plan and angular elevation of the school-house steps; scale, ¼ in. to the foot.

Prob. 14.—Draw plan and elevation of common spool.

Prob. 15.—Draw a plan and angular elevation of a flight of 4 steps, 12 in. by 3 ft. tread, and 7½ in. riser, with a rail 2 in. thick and 4 in. wide, supported by one baluster at each step 1½ in. from the front and end of the tread and 3 ft. high; scale, ½ in. to the foot.

Prob. 16.—Draw a plan and angular elevation of a cylinder lying on its side; diameter, 3 in.; length, 4½ in.; scale, ¼ in. to the inch.

Prob. 17.—Draw a plan and angular elevation of a hexagonal pyramid 21 in. long and 27 in. in diameter, lying on its side, and its axis making an angle of 45° with the elevation; scale, $\frac{1}{16}$ in. to the inch.

Prob. 18.—Draw plan and an angular elevation of a barrel lying on its side; diameter at the head, 18 in.; and at the bilge, 21 in.; length, 33 in.; scale, $\frac{1}{16}$ in. to the inch.

Prob. 19.—Draw plan and angular elevation of a common spool lying down with its axis making an angle of 30° with the elevation; scale, 2 in. to the inch.

Prob. 20.—Draw a plan and angular elevation of a spool with straight flanges (Fig. 92). Shank, 1 in. in diameter, 3 in. long inside the flanges; flanges, 1½ in. wide and 1 in. thick, and bore through the center $\frac{3}{4}$ in. in diameter; scale, 1½ in. to the inch.

Fig. 92.

Prob. 21.—Draw a plan and any angular elevation of a pulley 3 ft. in diameter; rim, 6 in. wide, 3 in. thick; hub, 9 in. in diameter, 6 in. long, and bore 3 in. in diameter; 4 spokes 1 in. × 5 in.; scale, 1 in. to the foot. (Fig. 93.)

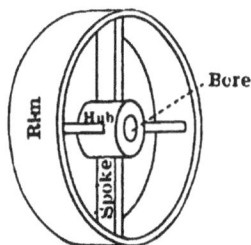

Fig. 93.

Prob. 22.—A piece of timber 13 in. square and 10 ft. long rests one end on the ground and the other end is 5 ft. from the ground, in the angle formed by two walls, the foot of the timber being equally distant from the walls forming the angle. Draw plan and elevation; scale, ½ in. to the foot. (Fig. 94.)

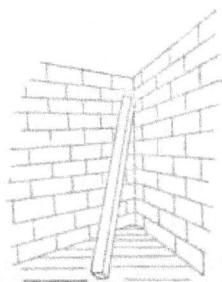

Fig. 94.

Prob. 23.—Draw a plan and front, side, and rear elevations of a common chair; scale, ½ in. to the ft.

Section V.—Two or more parallel elevations on the same base-line.

Prob. 1.—Draw the plan and parallel elevations of two adjacent sides of a box 6 in. long, 3 in. wide, 4½ in. high, ½ in. thick; scale, ⅛ in. to the inch.

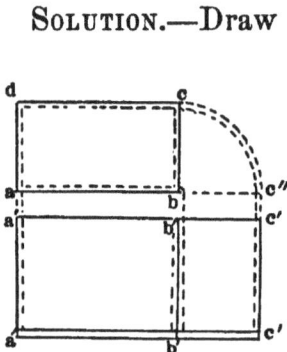

SOLUTION.—Draw the plan, *abcd,*' and draw the elevation, *a'a'b'b'*. Swing the side *cb* about the corner *b* until *c''b'* is in line with *ab*. Now lead *c''* down to the base-line *a'b*, and *b'b'c'c'* is the elevation of the side *CB*, and we have parallel elevations of two adjacent sides of an object on the same base-line. (Fig. 95.)

Fig. 95.

Prob. 2.—Draw the plan and elevations of two adjacent sides of a rectangular pyramid parallel to one side of the base. Base, 4 ft. × 7 ft., and pitch 8 ft.; scale, 1/16 in. to the foot.

SOLUTION.—Draw the plan *abcde*. Draw the end elevation, *a'b'e'*. Swing the side *bce* about the point *b*, into line with *ab*. Draw leaders from *e''* and *c''*, and construct the elevation *b'c'e'*. (Fig. 96.)

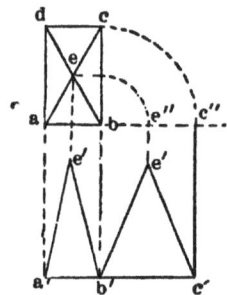

Fig. 96.

Prob. 3.—Draw the plan and parallel elevations of two sides of a box 6 in. wide, 5 in. long, 3 in. deep, and 1 in. thick. Bottom let in, ends laid on, and lid projecting 1 in. all around; scale, ¼ in. to the inch.

Prob. 4.—Draw the plan and parallel elevations of two sides of a rectangular frame-work 10 ft. long, 7 ft. wide, 9 ft. high, and made of timber 6 in. × 12 in., with the wider side of timbers vertical; scale, ⅜ in. to the foot.

Prob. 5.—Draw the plan and three elevations parallel to one of the sides of a triangular pyramid. Bases of sides, 3, 5, and 7 in., and pitch 4 in.; scale, ¼ in. to the inch.

Prob. 6.—Draw a plan and parallel elevations of three sides of a house 19 ft. sq., 16 ft. high, including foundation, 2 ft.; gable roof, 6 ft. pitch and 1 ft. projection. Door in middle of one end 4 ft. wide and 9 ft. high. Windows, 2 ft. 9 in. × 8 ft.,

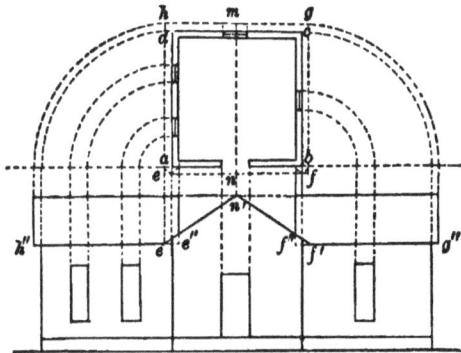

Fig. 97.—First Method.

2 ft. 3 in. from the floor, one in the middle of end opposite the door, one in the middle of the right-hand wall, and two dividing the remaining wall into three equal spaces on the outside; walls, 9 in. thick; scale, $\frac{1}{32}$ in. to the foot.

SOLUTION.—*First Method:* Draw the plan and con-

M. D —7.

struct the elevations as in the preceding problems,
and it will be seen that because the roof projects
beyond the body of the house, the gable and the
eaves of the front and side elevations lap over one
another (see first method, Fig. 97). This tends to
confound two different drawings, and in many cases
might lead to confusion and misunderstanding.
To avoid this, a method slightly different is adopted
by draughtsmen: .

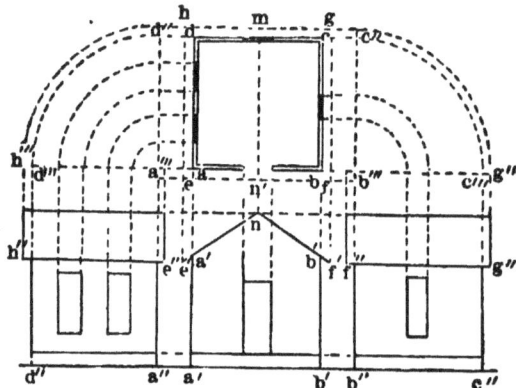

Fig. 98.—Second Method.

Second Method.—Draw the plan. Draw lines
parallel to and at any convenient distance from the
sides to be revolved, $a''d''$ and $b''c''$. Draw leaders
connecting these lines with the sides of the plan.
Revolve these lines as in the preceding method, and
construct the elevations. By this method, each elevation is entirely separate and distinct from the
others. (Fig. 98.)

Prob. 7.—Draw the plan and elevations of three
sides of a box 6 in. long, 5 in. wide, $4\frac{1}{2}$ in. high,
and $\frac{1}{2}$ in. thick. Mitered corners, bottom put on, .

and lid projecting 1 in. on all sides; scale, ½ in. to the foot.

Prob. 8.—Three timbers, one 4 ft. long, one 6 ft. long, and one 8 ft. long, and all 1 ft. square, are framed together perpendicular to each other at the middle. Draw plan and two elevations; scale, ½ in. to the foot. (Fig. 99.)

Fig. 99.

Prob. 9.—Draw plan and elevations of three sides of your school desk; scale, ½ in. to the foot.

Prob. 10.—Draw plan and elevations of four sides of your school-house; scale, ⅛ or $\frac{1}{16}$ in. to the foot.

CHAPTER IV.—Points in Inclined Surfaces.

Section I.—To find the elevation of a point in the edge of an inclined surface when the plan and the elevation of the surface and the plan of the point are given.

Prob. 1.—The plan of an inclined surface is a rectangle 7 ft. × 8 ft. The elevations of the 8 ft. edges are horizontal, and the elevations of the 7 ft. edges .have a pitch of 6 ft. In the right hand inclined edge of this surface is a point, and the plan of this point is in the right hand shorter edge of the plan, 4 ft. from the plan of the lower edge. It is required to describe this point and the surface by a plan and elevation drawing; scale, $\frac{1}{16}$ in. to the foot.

SOLUTION.—As the plan of the given inclined surface is a rectangle 7 ft. × 8 ft., and
the scale is $\frac{1}{16}$ in. to the foot, draw a
rectangle, *abcd*, $\frac{7}{16} \times \frac{8}{16}$ in. From
this plan draw leaders, and construct
an angular elevation, *a'b'c'd'*, having
a pitch of $\frac{6}{16}$ in., *ed'*. This fully describes the given inclined surface.
As the plan of the given point is in
the right hand side of the plan of
the surface it will be in *bc*. As it is 4 ft. from the
lower horizontal edge, it will be at *p*, $\frac{4}{16}$ in., from
ab on *bc*. It is required to find its elevation. As
the plan of the point is on *bc*, its elevation must be
on the elevation of $BC = b'c'$. It must also be on a
leader drawn from *p* to intersect *b'c'* in *p'*. The
point *P*, in the edge of the inclined surface, is now
completely described by its plan and elevation *p* and
p'. (Fig. 100.)

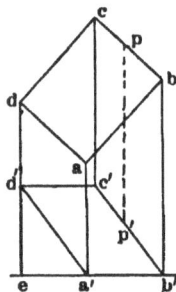

Prob. 2.—An inclined surface has a rectangular
plan, $4\frac{1}{2}$ in. × $7\frac{3}{4}$ in.; the elevations of the shorter
edges are horizontal, and the pitch of the other
edges is $3\frac{1}{8}$ in. In the left-hand inclined edge is a
point, and the plan of this point is $2\frac{1}{4}$ in. from the
plan of the upper edge of the surface. Draw a
plan and elevation that will completely describe the
given point and surface; scale, $\frac{1}{2}$ in. to the inch.

Prob. 3.—A wedge stands on
its base, which is 9 ft. × 18 ft.,
and the pitch is 8 ft. In each
of the edges of one of its sides
is a point, *a*, *b*, *c*, and *d*. The
plan of *a* is on the plan of the ridge 4 ft. from the

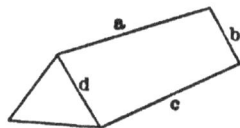

Fig. 100.

Fig. 101.

right-hand end; the plan of *b* is on the plan of the right-hand edge, and 2½ ft. from the plan of one of the longer edges of the base; the plan of the point *c*, in the longer side of the base, is 8 ft. from the left-hand end; the plan of the point *d* is in the plan of the left-hand edge and 4 ft. from the plan of the ridge. Draw plan and elevation; scale, ⅛ in. to the foot. (Fig. 101.)

Prob. 4.—A rectangular pyramid, base 12 in. × 9 in., and pitch 18 in., stands on its base and has a point in each of the edges of one of its larger triangular sides, *a*, *b*, and *c*. (Fig. 102.) The plan of *a* is on the plan of the left-hand edge of the side, 4 in. from the plan of the base; the plan of the point *c* is in the middle of the plan of one of the longer sides of

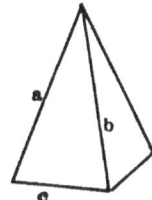

Fig. 102.

the base; and the plan of *b* is in the plan of the right-hand edge 3¾ in. from the plan of the apex of the pyramid. Draw plan and elevation; scale, 3/16 in. to the inch.

Prob. 5.—Draw plan and elevation of an oblique triangular pyramid; sides of base, 3 ft.; pitch, 4½ ft.; and the plan of the apex on a line bisecting one of the angles of the base, and 5 ft. from it; scale, 1 in. to the foot.

Rule.—Draw the plan and elevation of the given inclined surface, and locate the given plan of the required point. From this plan of the point, draw a leader to intersect the corresponding edge of the elevation, and this point of intersection will be the required elevation of the given point.

Section II.—*To find the elevation of a line that joins two edges of an inclined surface when the plan and elevation of the surface and the plan of the line are given.*

Prob. 1.—The plan of an inclined surface is a rectangle 25 ft. × 32 ft.; the elevation of the 32 ft. edges are horizontal, and the pitch of the other edges is 22 ft., and the plan of a line crossing this surface has one end in the left-hand edge of the plan of the surface 5 ft. from the plan of the lower horizontal edge, and the other end is in the plan of the upper edge of the surface, 6 ft. from the right-hand edge of the plan. It is required to draw the plan and elevation of the surface and the line; scale, $\frac{1}{64}$ in. to the foot.

SOLUTION.—Draw a rectangle, *abcd*, $\frac{25}{64}$ in. × $\frac{32}{64}$ in.,

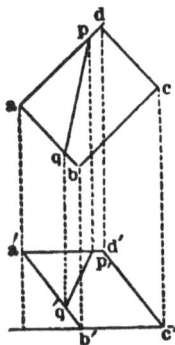

Fig. 103.

for the plan of the given surface. From this draw the angular elevation, *a'b'c'd'*, with the longer edges horizontal, and with a pitch of $\frac{22}{64}$ in. Locate the ends, *p* and *q*, of the plan of the given line *pq*, *p* on *ad*, $\frac{6}{32}$ in. from *dc*, and *q* on *ab*, $\frac{5}{32}$ in. from *bc*. Join *p* and *q*, and *pq* is the plan of the given line. Find the elevations of *p* and *q* in *p'* and *q'*, and join them, and *p'q'* is the elevation of the given line.

Prob. 2.—One side of a gable roof is 16 ft. long, 10 ft. span, and 8 ft. pitch (the plan a rectangle 16 ft. × 10 ft.). It is crossed by three lines: one is 3 ft. from the eaves, and parallel to it; one is $4\frac{1}{2}$ ft.

from the right-hand gable end, and parallel to it, and one is oblique; one end of the plan of the oblique line is in the plan of the upper edge, 5 ft. from the right-hand end, and the other end is in the plan of the eave, 7 ft. from the left-hand end. Draw the plan and elevation of the roof and lines; scale, ⅛ in. to the foot.

Prob. 3.—An equilateral triangular pyramid measures 33 in. at the base of each side, and has a pitch of 48 in.; it is cut by a plane which cuts each side in a straight line. There are three sides, with one straight line in each side, and these lines meet on the hip (edges of the sides). The plans of the points on the hips are 4 in., 9 in., and 15 in. from the plan of the apex. Draw the plan and elevation; scale, $\frac{1}{16}$ in. to the foot.

Prob. 4.—A rectangular pyramid, 3 ft. square at the base and 9 ft. pitch, is cut by a plane. The plan of the points of intersection with the hips are 15 in., 12 in., 9 in., and 12 in. from the plan of the apex. Draw the plan and elevation; scale, 1 in. to the foot.

Rule.—Draw the plan and elevation of the given inclined surface, and the plan of the given line on the surface. From the ends of the plan of the given line, draw leaders to intersect the elevations of the edges of the surface in which the ends of the given line rest. These points of intersection are the elevations of the ends of the given line. Join these points of intersection by a right line, and this line will be the required elevation of the given line.

Section III.—To find the elevation of a point in an inclined surface when the plan and elevation of the surface and the plan of the point are given.

Prob. 1.—In one side of a gable roof 27 ft. long,

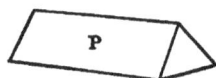

Fig. 104.

14 ft. span, and 6 ft. pitch, is a hole, and the plan of this hole is 13 ft. from the plan of the left-hand gable end and 4 ft. from the plan of the eave. It is required to find the elevation of the given point; scale, $\frac{1}{32}$ in. to the foot. (Fig. 104.)

SOLUTION.—Draw the plan *abcd*, and locate the plan of the given point, *p*. Through *p* draw any line, *ef*, crossing the plan of the side of the roof. Find the elevation of the roof and the line *cf*. From *p* draw a leader to intersect *e'f'* in *p'*, and *p'* is the required elevation of the given point. (Fig. 105.)

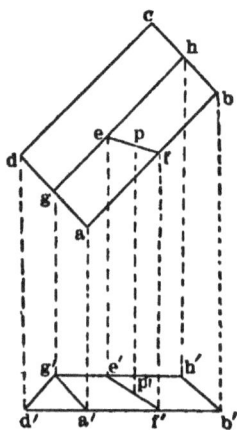

Fig. 105.

Prob. 2.—Draw the plan and elevation of a rectangular pyramid, 9 in. × 12 in. base, and 8 in. pitch, pierced by a line parallel to the axis; the plan of the point where this line pierces one of the larger of the inclined surfaces of the pyramid is 1½ in. from the edge of the plan of the base, and 5 in. from the left-hand edge of the plan of the face that it pierces; scale, ⅛ in. to the foot.

Prob. 3.—A coppersmith is required to construct an equilateral triangular prism, each face of which is 8 in. wide and measures 13 in. parallel to the axis. Through this prism is to pass a fine straight wire, perpendicular to one of its faces, which it pierces 2 in. from one edge and 4 in. from the end. Draw plan and elevation of the prism and the holes for the wire; scale, ⅛ in. to the inch.

Rule.—Draw the plan and angular elevation of the surface; locate the given plan of the point, and through it draw any line crossing the plan of the surface. Find the elevation of a line on the given surface of which this line is the plan. Draw a leader from the plan of the point to intersect this line, and the point of intersection will be the required elevation of the point.

Section IV.—To find the elevation of a point in an inclined surface by parallel elevations on the same base when the plan of the point and the plan and elevation of the surface are given.

Prob. 1.—Draw the plan and parallel elevations of the south and east sides of a gable-roofed house, standing with its longer sides north and south; 16 ft. wide, 20 ft. long, and 10 ft. high; 1 ft. foundation; roof pitch, 9 ft., and projection, 1 ft. Plan of chimney, 2 ft. square, 3 ft. 6 in. from the east wall, and 11 ft. from the north wall; chimney projection above the roof, 6 ft.; scale, $\frac{1}{24}$ in. to the foot.

SOLUTION.—Draw the plan, *abcdmnef*, and the end elevation, *a'b'*. This gives us the height of the point where the chimney comes out of the roof. Revolve the side *bc*, and construct a side elevation of the house and chimney. The elevation of the point where the chimney will come through the roof is located in *e'''f'''*.

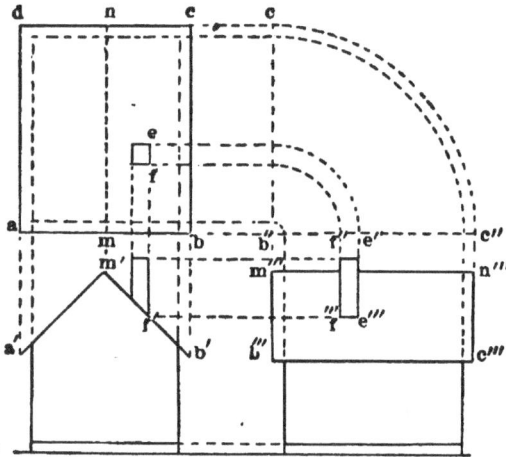

Fig. 106.

Prob. 2.—A solid rectangular pyramid, 4 in. × 6 in. base and 9 in. pitch, is pierced by a one inch square hole, perpendicular to its base, and cutting it midway between its shorter edges, and midway between the plan of the apex and the plan of the base; scale, ½ in. to the inch.

Prob. 3.—Draw a plan and angular elevation of a hexagonal pyramid, 12 in. diameter × 18 in. long, lying on its side, and pierced by a vertical line which passes through it 4½ in. from the axis and 7 in. from the end; scale, ¼ in. to the inch.

Prob. 4.—The sides of the base of a triangular pyramid are 1¼ in., 1½ in., and 2 in., and the pitch is 4½ in.; each of its sides is pierced by a vertical line. The plan of one of these lines is ¼ in. from the base and ⅝ in. from the right-hand edge of the plan of the longest face; the plan of the second line is ⅛ in. from the shortest base, and 1 in. from the left-hand edge of the plan of the smallest face; and the plan of the third is ₁³₆ in. from the base and ⅝ in. from the plan of the right-hand edge of the third face. Draw the plan and the elevation; half size.

Rule.—Draw the plan of the given surface, and locate the given plan of the required point. From this plan construct an end elevation of the given surface and point, and from these construct a parallel side elevation on the same base-line as the end elevation. The three drawings so constructed will completely describe the given surface and the required point.

CHAPTER V.—THE DESCRIPTION OF POINTS IN A SURFACE OF REGULAR CURVATURE.

A Surface of Regular Curvature follows a fixed law.

REMARK.—A sphere has a surface of regular curvature, because, having any point on its surface and the center of the sphere given, its entire surface may be exactly determined and described.

An egg has a surface of irregular curvature, because, having any number of points given, it would still be impossible to determine exactly and describe the remaining points on its surface.

*Section I.—To find a 'point on a regular curved sur-
face when the plan and elevation of the surface
and the plan of the point are given.*

Prob. 1.—A hemisphere 2 in. in diameter has a
point on its surface, and the plan of this point is ⅝
in. from the center of the plan of the sphere.
Draw the plan and elevation of the hemisphere
and point; scale, ½ full size.

SOLUTION.—Draw the plan and elevation of the
hemisphere *ab* and *a'b'd'*. (Fig.
107.) As the surface of a
sphere may be regarded as
consisting of any number of
parallel circles increasing and
diminishing in diameter according to a fixed law, every
point on the surface of a
hemisphere may be regarded
as located on a circle parallel
to its base. Locate the point
p ⅝ in. from the center of
the plan, and through it describe a circle, *ef*, parallel
to *ab*. This circle, *ef*, is the
plan of a circle on the surface
of the hemisphere. To find
the elevation of *ef*, draw leaders tangent to it, to intersect
the outline of the elevation in *c'* and *f'*. Join *e'*
and *f'*, and *e'f'* is the elevation of a circle on the

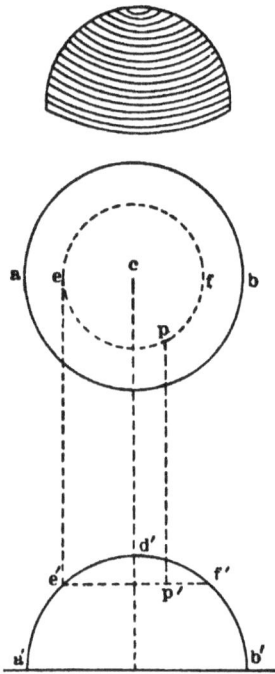

Fig. 107.

surface of the hemisphere which contains the given point. Draw the leader, pp', to intersect $e'f'$, and p' is the required elevation of the given point, which is now completely described.

Prob. 2.—Draw the plan and elevation of a cone; base, 46 ft. in diameter; and pitch, 63 ft., with a point on its surface, of which the plan is 13 ft. from the center of the plan of the cone; scale, $\frac{1}{64}$ in. to the foot.

SOLUTION.—Draw the plan and elevation of the cone, abc and $a'b'c'$. As the surface of the cone may be regarded as made up of a series of circles parallel to each other, and increasing and diminishing according to a fixed law, the point on the surface of this cone may be considered as being on one of these circles, the plan of which is 13 ft. radius.

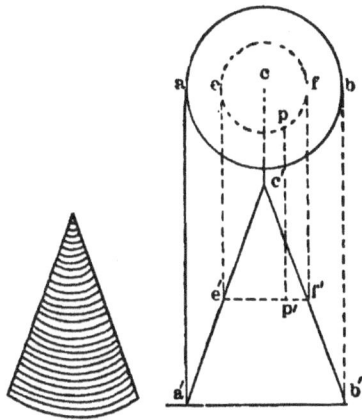

Fig. 108.

Draw the plan of this circle, ef, and from this draw leaders tangent to it to meet the elevation in e' and f'. Draw $e'f'$, and this will be the elevation of the circle containing the elevation of the point given. Locate the plan of the point p, and from it draw a leader to intersect $e'f'$ in p'; p' is the elevation of the given point, and the point and the cone are completely described.

Prob. 3.—Draw the plan and elevation of a hemisphere 10 ft. in diameter, with a point on its surface, the plan of which is 3 ft. from the center of the plan of the hemisphere; scale, $\frac{1}{8}$ in. to the foot.

Prob. 4.—Draw the plan and elevation of a cone, 1 in. radius at the base and $2\frac{1}{2}$ in. pitch, with a point on its surface, the plan of which is $\frac{5}{8}$ in. from the center of the plan of the cone; scale, half size.

Prob. 5.—Draw the plan and elevation of a sphere 5 ft. in diameter, having two points on its surface, one above and one below the center. The plan of the point above the center is 1 ft., and the plan of the other is 9 in., from the plan of the center of the sphere; scale, half size.

Prob. 6.—A cone, 10 ft. in diameter at the base and 15 ft. pitch, is pierced by four horizontal lines 12 ft. long. The plan of the point where the surface of the cone is pierced by one of the lines is 1 ft. from the center of the plan of the cone; the plan of another is 2 ft. from the plan of the center; the plan of the third is 3 ft. from the plan of the center, and the plan of the fourth is 4 ft. from the plan of the center. All the lines project equally, and their plans form equal angles with each other, and meet on the vertical axis of the sphere. Draw the plan and elevation; scale, $\frac{1}{8}$ in. to the foot.

Prob. 7.—A hemisphere, 33 ft. radius, is pierced in each quadrant by a line that is 20 ft. in length in the plan, and passes through the center. The plan of a point where one of these lines pierces the surface of the hemisphere is 5 ft. from the center of the plan and 6 ft. from the right-hand side of the first quadrant; the plan of another point is 14 ft.

from the center and 7 ft. from the left-hand side of second quadrant; the plan of the third point is 9 ft. from the center of the plan and 3½ ft. from the right-hand side of third quadrant; and the plan of the fourth point is 11 ft. from the center and 4 ft. from the left-hand side of fourth quadrant. Draw the plan and elevation; scale, $\frac{1}{16}$ in. to the foot.

Section II.—To find the plan of a point on a surface of regular curvature when the plan and elevation of the surface and the elevation of the point are given.

Prob. 1.—Draw the plan and elevation of a hemisphere 3 in. in diameter, and having a point on its surface, of which the elevation is 1¼ in. from the base-line and 1⅛ in. measured horizontally from the right-hand edge of the elevation of the hemisphere; scale, ¼ full size. (Fig. 109.)

SOLUTION.—Draw the plan and elevation of the hemisphere, *abd* and *a'b'g*, and locate the point *p'* in the elevation. Through the point *p'* draw *e'f'* parallel to the base, and it will be the elevation of a circle passing through the given point on the surface of the hemisphere. From *e'* and *f'* draw leaders, and with *c* as a center describe the plan of the circle, *ef*, tangent to the leaders. Draw the leader *pp'* to meet this circle, *ef; p* is the required

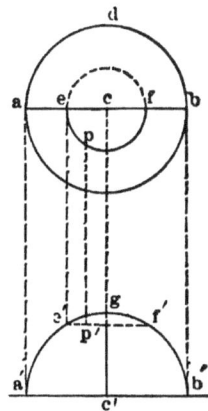

Fig. 109.

plan of the given point, and the hemisphere and the point on its surface are completely described.

Prob. 2.—A hemisphere 25 ft. in diameter has a point on its surface, the elevation of which is 2 ft. high and 6 ft. measured horizontally to the left-hand of the vertical axis. Draw the plan and elevation of hemisphere and point; scale, ⅛ in. to the foot.

Prob. 3.—A cone, 3 in. in diameter at the base and 4½ in. pitch, is pierced by a vertical line at a point on the surface of the cone one third of its height above the base. Draw the plan and elevation of the cone and line; scale, ½ in. to the inch.

Prob. 4.—A sphere 12 in. in diameter is pierced by two sets of three horizontal lines, each 15 in. in length, and the lines of each set form equal angles and meet on the vertical diameter of the sphere. One set of lines is 2 in. above, and the other is 4½ in. below, the center of the sphere. Draw the plan and elevation of the sphere and the lines; scale, ¼ in. to the inch.

Prob. 5.—A cone, 100 ft. in diameter at the base and 150 ft. pitch, is pierced by five lines, passing through the center of the base and piercing the surface of the cone. The elevations of the points on the surface of the cone arc located as follows: 24 ft. high, and 40 ft. to the left-hand of the axis; 12 ft. high, and 10 ft. to the right-hand of the axis; 40 ft. high, and 36 ft. to the left-hand of the axis on one side; 90 ft. high, and in the middle; and 75 ft. high, 28 ft. to the left-hand of the axis on the other side. Draw the plan and elevation of the cone and the lines; scale, $\frac{1}{64}$ in. to the foot.

CHAPTER VI.—The Intersection of Horizontal Lines with Inclined Surfaces.

Prob. 1.—Draw the plan and elevation of a house 16 ft. wide, 20 ft. long, and 18 ft. high, including foundation, 2 ft.; gable roof, 10 ft. pitch, 16 ft. span, and projecting 1 ft.; dormer window, with gable roof in middle of side of roof, 4 ft. wide, 8 ft. high to the ridge of the roof, 2 ft. pitch, projecting 6 in., and the face flush with the wall of the house; scale, $\frac{1}{24}$ in. to the foot.

Solution.—Draw the end elevation of a house and the side elevation of a dormer window, $a'b'o'p'm'a'$. From this construct the plan, and find p,

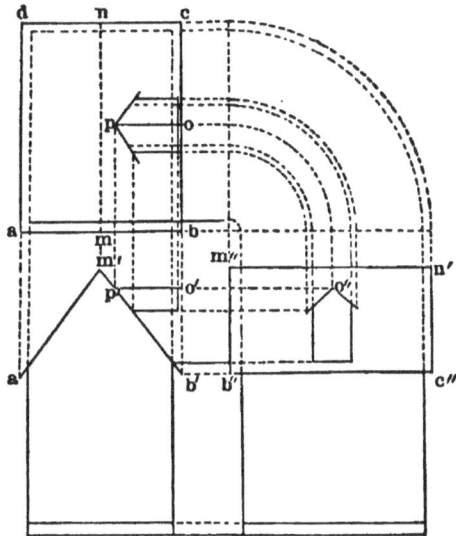

Fig. 110.

which is the plan of the point where the ridge of the dormer window, op, will meet the roof. From these construct the side elevation. (Fig. 110.)

Prob. 2.—A house is 18 ft. × 28 ft. × 20 ft. high, with a gable roof, 9 ft. pitch, and a wing 16 ft. × 3 ft. projection, and 16 ft. high; eaves and gable project 1 ft. (Fig. 111.) Problem: to find where the ridge

Fig. 111.

M. D.—8.

of the wing will meet the roof of main house; scale, ⅛ in. to the foot.

Prob. 3.—A house is 24 ft. square, 18 ft. high, with a gable roof 8 ft. pitch; it has an extension 12 ft. square and 12 ft. high, projecting 4 ft. beyond, and extending 8 ft. along the side of the main house. Make a working drawing; scale, ⅛ in. to the foot.

Fig. 112.

Def.—A working drawing is a complete description of all the dimensions of an object by plans and elevations.

Prob. 4.—A barn is 18 ft. × 24 ft., 16 ft. high, gable roof ¾ pitch (¾ span), with a dormer window 5 ft. wide and 5 ft. high, having a gable roof, ¾ pitch, 7 ft. from the end of the barn. Make a working drawing; scale, ⅛ in. to the foot. (Fig. 113.)

Fig. 113.

Prob. 5.—A house is 24 ft. × 36 ft., 20 ft. high, ½ pitch, gable roof. There is a projection 3 ft. × 18 ft., 6 ft. from right-hand end of the longer side, having a gable roof ½ pitch; and a dormer window, 5 ft. wide and ½ pitch, 4 ft. from the left-hand end of the roof and 6 ft. from the eaves. Make a working drawing; scale, ⅛ in. to the foot. (Fig. 114.)

Fig. 114.

Rule.—Draw the end elevation of the inclined surface and the side elevation of the given line.

From this, draw the plan of the surface and line. From these draw the side elevation of the surface and the end elevation of the line. By the plan and elevations so drawn, the line will be completely described.

MISCELLANEOUS PROBLEMS.

Prob. 1.—A house is 24 ft. × 40 ft., 24 ft. high, and ½ pitch gable roof (12 ft.). In the middle of one of the longer sides is a rectangular oriel window, 10 ft. wide, 9 ft. high, and projecting 3 ft.; with gable roof one half pitch (5 ft.), and eaves continuous with eaves of main house. There is a hexagonal oriel window at one corner, 8 ft. in diameter and 9 ft. high, showing 5 faces; this window has a one half pitch (4 ft.) gable roof, and its eaves are continuous with the eaves of the

Fig. 115.

main house. Make a ground plan, a second story, and a roof plan, and an end and two side elevations; scale, ⅛ in. to the foot. (Fig. 115.)

Fig. 116.

Prob. 2.—A house is 36 ft. in diameter; it is 15 ft. high, and has a hemispherical roof; there is a wing 14 ft. × 16 ft., 15 ft. high, with a half pitch roof. Make a plan, a side, and an end elevation; scale, ⅛ in. to the foot. (Fig. 116.)

Prob. 3.—A house is 24 ft. square, 18 ft. high, and has a hip roof, ¾ pitch; it has a wing 14 ft. square and 18 ft. high, 2 ft. from the left-hand corner of the house with a half pitch roof; a dormer window 4 ft. wide and 3 ft. high, with a lean-to roof, 3 ft. pitch, and the sill 3 ft. from the eaves. Draw a ground plan, a second floor plan, a roof plan, and a front and a side elevation; scale, ⅛ in. to the foot. (Fig. 117.)

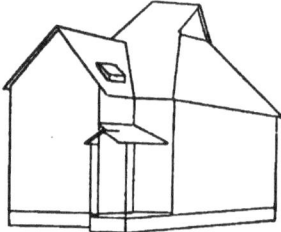

Fig. 117.

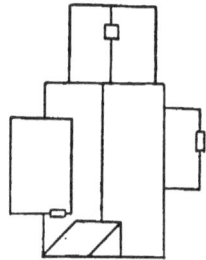

Fig. 118.

Prob. 4.—Draw plans and elevations of a house with a roof similar to that described by Fig. 118.

CHAPTER VII.—Conic Sections.

Prob. 1.—Draw the plan and elevation of a frustum of a cone 5 in. in diameter at the base, 8 in. pitch, and cut off at the top by a plane cutting the axis 4½ in. from the base at an angle of 45°; scale, ¼ in. to the inch.

Solution.—Draw the plan and elevation of the cone, abc and $a'b'c'$, and the end elevation, $m'n'$, of the cutting plane, and $a'b'n'm'$ is the elevation of the required frustum. It is required now to find the plan of the frustum where it is cut off by the plane $m'n'$. Draw the elevations and plans of any number of circles, 1–5, on the surface of the cone that will be cut by the plane $m'n'$ and the points 1', 2', 3', 4', 5, m', and n' will be the elevations of ten points in the top of the frustum. Find the plans of these points, and join them by a curved line which will describe the plan of the top of the required frustum. (Figs. 119, 120, 121.)

Fig. 119.   Fig. 120.

Prob. 2.—Draw a plan and elevation of a cone 13 ft. in diameter at the base, 25 ft. pitch, and cut by a plane making an angle of 60° with the axis, 15 ft. from the base; scale, $\frac{3}{16}$ in. to the foot.

Prob. 3.—Draw a plan and elevation of a frustum of a cone $4\frac{1}{2}$ in. in diameter at the base, 5 in. pitch, and the plane of the top making an angle of 45° with the axis, $2\frac{1}{2}$ in. from the base; scale, $\frac{3}{4}$ in. to the foot.

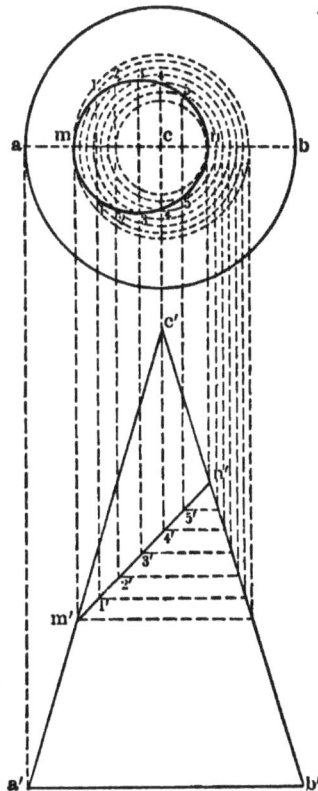

Fig. 121.

Prob. 4.—Draw a plan and elevation of a frustum of a rectangular pyramid, base 63 ft. square, pitch 105 ft., angle of the top 90° with one of the sides, and length of axis 80 ft.; scale, $\frac{1}{82}$ in. to the foot.

Prob. 5.—Draw the plan and elevation of a frustum of a hexagonal pyramid 2 in. in diameter at the base, 4 in. pitch, angle of the top 45° with one of the hips (angles), and the axis 3 in. in length; scale, full size.

Prob. 6.—Draw the plan and elevation of a frustum of an oblique cone, 23 ft. in diameter at the base; angle of the axis with base, 60°; length of axis, 15 ft.; top, 4 ft. radius, and parallel to base; scale, $\frac{1}{8}$ in. to the foot.

Prob. 7.—Draw the plan and elevation of a frustum of an oblique cone, 10 ft. in diameter at the base, axis inclined 60° from the vertical and 5 ft. long, upper face perpendicular to axis, and 5 ft. wide in its longest diameter; scale, $\frac{1}{4}$ in. to the foot.

CHAPTER VIII.—The Helix.

A **Helix** is a spiral line that continually advances in the direction of a straight line called its axis.

If the spiral line ascends and passes from left to right in front of the axis, when the axis is vertical, the helix is said to be *right-hand.*

If the spiral line ascends and passes from right to left in front of the axis, when the axis is vertical, the helix is said to be *left-hand.*

The **Pitch** of a helix is the distance it advances in making one revolution, and is measured parallel to the axis.

Section I.—Helical Lines.

Prob. 1.—Draw the plan and elevation of a right-hand helical line 1 in. in diameter, 2½ in. long, and 1½ in. pitch.

SOLUTION.—As the required helix is 1 in. in diameter and 2½ in. in length, draw the plan and elevation of a cylinder having those dimensions, *acb* and *a'c'c''a''* (Fig. 122).

Since the pitch is 1½ in., draw the line *a'''c'''* 1½ in. from *a'c'*.

As this helix advances from *a'* to *a'''* in making one revolution, it will advance $\frac{1}{12}$ of that distance in making $\frac{1}{12}$ of one revolution.

This being a right-hand helix, it will pass from left to right. Begin at *a* and. *a'*, and divide the plan, *abc*, and the pitch, *a'a'''*, into 12 equal divisions, 1-2 to 12. Through these division points in the elevation, draw dotted horizontal lines, and from the division points in the plan draw leading lines to intersect them. Join these points

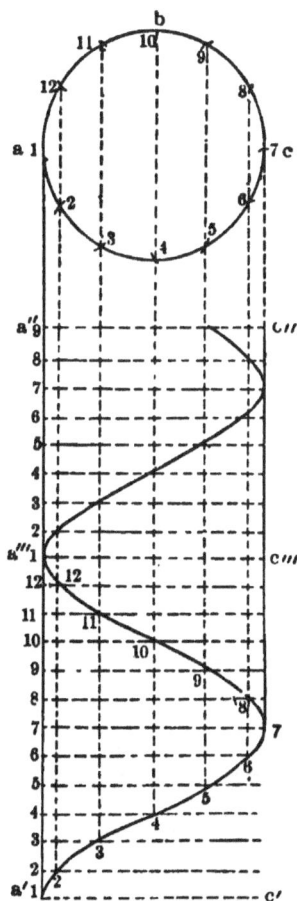

Fig. 122.

of intersection by a curved line. This line will be
one pitch (1½ in.) of the required helix. The remaining portion (1 in.) may be described in the
same way.

> NOTE.—To describe a left-hand helix, the line would pass
> from right to left instead of from left to right, as in the last
> problem; that is, in describing a left-hand helix, *a* and *c* would
> exchange places.

Prob. 2.—Make a constructive drawing of a right-
hand helical line, 1 ft. pitch, 1 ft. diameter, and 2
ft. long; scale, 2 in. to the foot.

Prob. 3.—Make a constructive drawing of a left-
hand helical line, 6 in. pitch, 3 in. diameter, and 9
in. long; scale, ½ in. to the foot.

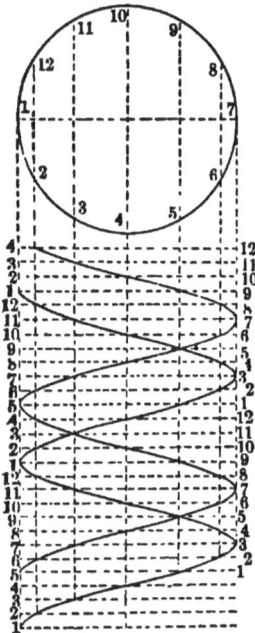

Fig. 123.

Section II.—Helical Bands.

A **Helical Band** is the space
between two similar and parallel helices.

Prob. 1.—Make a constructive
drawing for a right-hand helical band ¼ in. wide, 1 in. in diameter, ¾ in. pitch, and 1¹¹⁄₁₆ in.
long.

SOLUTION.— Draw plan and
elevation of a cylinder 1 in. in
diameter and 1¹¹⁄₁₆ in. long. (Fig.
123.)   Construct a right-hand
helix, ¾ in. pitch. Beginning
¼ in. from the bottom of the
elevation, draw another right-hand helix having the

same pitch as the first. The space between these helices will be the helical band required in the problem.

Prob. 2.—Make a constructive drawing for a right-hand helical band ¼ in. wide, 2 in. long, 2 in. in diameter, 1¼ in. pitch.

Prob. 3.—Make a constructive drawing for a left-hand helical band ¾ in. wide, 2½ in. long, 1⅜ in. radius, and 1⅜ in. pitch.

Prob. 4.—Make a constructive drawing for a left-hand helical band 10 ft. wide, 43 ft. long, 28 ft. diameter and 36 ft. pitch; scale, $\frac{1}{16}$ in. to the foot.

Prob. 5.—Make a constructive drawing for a right-hand and a left-hand helical band, ¼ in. wide, crossing each other. Diameter 2 in., length 2 in., pitch 2 in.

Section III.—Helical Flanges.

A **Helical Flange** is a surface that is between two parallel helices, and is perpendicular to the axis of the helix.

Prob. 1.—Make a constructive drawing for a right-hand helical flange ¼ in. wide, 1 in. outer diameter, 1$\frac{5}{16}$ in. long, and ¾ in. pitch.

SOLUTION.—Draw plan and elevation for a right-hand helix, 1 in. diameter, 1$\frac{5}{16}$ in. long, and ¾ in. pitch, 1-4-7-10-12

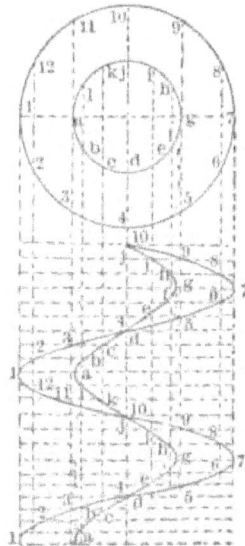

Fig. 124.

(Fig. 124). Within this helix, and on the same axis, draw the plan and elevation of another right-

M. D.—9.

hand helix $\frac{1}{2}$ in. diameter, $1\frac{5}{16}$ in. long, and $\frac{3}{4}$ in. pitch, *a-d-g-j.* The space between these helices rep, resents the flange required in the problem.

Prob. 2.—Make a constructive drawing for a right-hand helical flange $\frac{1}{2}$ in. wide, 2 in. inner diameter, 3 in. pitch, and 4 in. long.

Prob. 3.—Make a constructive drawing for a left-hand helical flange $2\frac{1}{2}$ in. in diameter, $1\frac{1}{2}$ in. pitch, 1 in. wide, and 3 in. long.

Prob. 4.—A cylinder 2 in. in diameter and $4\frac{1}{2}$ in. long, has on its surface a right-hand helical flange $1\frac{1}{2}$ in. pitch and $\frac{1}{8}$ in. wide. Make a constructive drawing; scale, $\frac{1}{2}$ in. to the inch.

Section IV.—Angular Helical Projections.

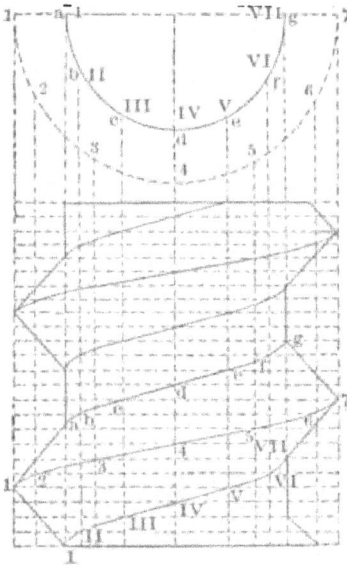

Fig. 125.

Prob. 1.—Make a constructive drawing for a screw $1\frac{1}{2}$ in. in diameter, $1\frac{1}{2}$ in. long, $\frac{3}{4}$ in. pitch; V thread, $\frac{1}{4}$ in. deep and $\frac{1}{2}$ in. wide.

SOLUTION.—Draw the plan and elevation of a helical band, $\frac{1}{2}$ in. wide, $\frac{3}{4}$ in. pitch, and 1 in. diameter and $1\frac{1}{2}$ in. long, *I-VI* and *a-g* (Fig. 125). Draw the helix 1-7, beginning $\frac{4}{16}$ in. from bottom of elevation, $1\frac{1}{2}$ in. diameter, and $\frac{3}{4}$ in. pitch. Connect these helices by straight lines, and the screw is described.

Prob. 2.—Make a constructive drawing for a right-hand V thread screw, 2½ in. long, 4¼ in. diameter, 2 in. pitch, and threads ⅞ in. deep and 1¾ in. wide.

Prob. 3.—Make a constructive drawing for a left-hand V thread screw 3 in. long, 1½ in. in diameter, 1⅓ in. pitch; threads 11/16 in. deep and 1½ in. wide.

Prob. 4.—Make a constructive drawing for a 3½ in. right-hand V thread screw, 1 in. long, 1¼ in. pitch; threads meet at bottom of groove, the depth of which is equal to ½ the pitch.

Prob. 5.—Make a constructive drawing for a left-hand V thread screw 4 in. in diameter, 3 in. long; angle of thread 87½°, and pitch 1½ in.

Section V.— Rectangular Helical Projections.

Prob. 1.—Make a constructive drawing for a right-hand, rectangular, helical projection 1½ in. in diameter, 1⅜ in. pitch, and ¼ in. square, on a cylinder 1⅛ in. long.

SOLUTION.—Draw the band, 1-7 and *I-VII* (Fig. 126), ¼ in. wide, 1½ in. in diameter, and 1⅜ in. pitch. Draw the cylinder 1⅛ in. long and 1 in. in diameter. Draw the parts of the helices,

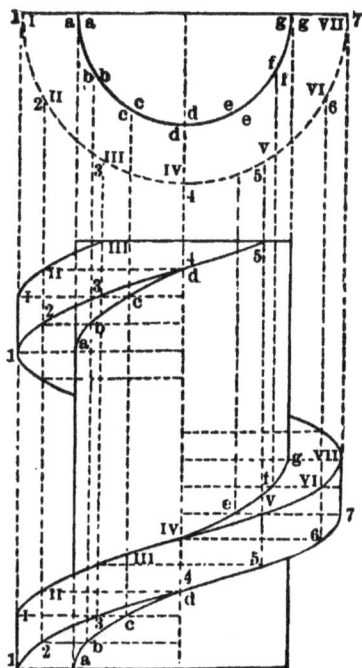

Fig. 126.

abcd and *efg*, that show at the upper and lower sides of the helical projection. Draw the exposed parts of the cylinder, and the conditions of the problem are completely satisfied.

Note.—The helical projection described in Fig. 126 is called a *square thread.*

Prob. 2.—Make a constructive drawing for a left-hand, square thread screw, 3 in. long, 1½ in. pitch, 2½ in. diameter, and thread ¾ in. square.

Prob. 3.—Make a constructive drawing for a right-hand, square thread screw, 16 in. long, 24 in. in diameter, 6 in. pitch; thread, ½ of the pitch; scale, 1/16 in. to the inch. (Fig. 127.)

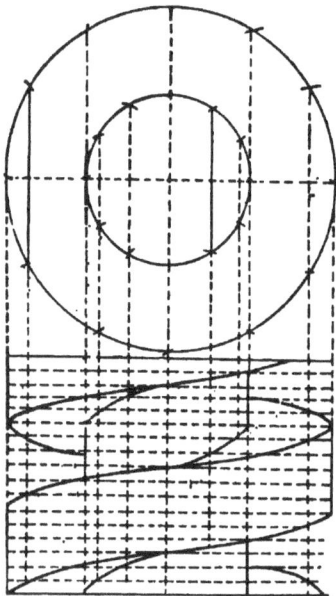
Fig. 127.

Prob. 4.—Make a constructive drawing for left-hand, square thread screw 2 in. long, 1½ in. diameter, 1½ in. pitch.

Prob. 5.—Make a constructive drawing from the object for a square thread screw with head, nut, and washer, full size.

Prob. 6.—Make a constructive drawing for a flight of 16 winding stairs, inner radius 1 ft. and outer radius 5½ ft.; riser, 7½ in. Plan of stairs a complete circle, and each step a single stone block setting over 1½ in. on the step below it; scale, ¼ in. to the foot.

Prob. 7.—Make a constructive drawing for a right-hand hand-rail 6 in. wide, 3 in. thick, 10 ft. pitch, 5 ft. radius, and starting 3 ft. from the floor.

Prob. 8.—Make a constructive drawing from the object for a flight of stairs with winders and balustrade.

Prob. 9.—Make a constructive drawing for a circular staircase tower 25 ft. outside diameter, 39 ft. high, and wall 18 in. thick, pierced by 16 rectangular windows, 2 ft. wide and 8 ft. high, equally distant apart, and arranged in a helical line, making one revolution of the tower, and having a 20 ft. pitch, beginning 4 ft. from the ground; scale, $\frac{1}{4}$ in. to the foot.

CHAPTER IX.—Section Drawing.

A **Section Drawing** describes the dimensions of an object where it is cut by a given plane.

Prob. 1.—A cylinder 10 ft. long, 10 ft. in diameter, and 3 ft. bore, and standing with its axis vertical, is cut by three planes; one of these planes is horizontal, and cuts the cylinder 5 ft. from the bottom; one plane is vertical, and passes through the axis; and one is oblique, and makes an angle of 45° with the axis, cutting it 3 ft. from the base. Make a constructive drawing for the cylinder and each section; scale, $\frac{1}{16}$ in. to the foot.

Solution.—Draw the plan and elevation of the cylinder, *ab* and *a'b'a'b'*. Draw the line *EF* for the elevation of the horizontal cutting plane, and perpendicular to it, draw leaders from the points where it crosses the different parts of the elevation, and between these leaders construct the cross section drawing $e'f'$. (Fig. 128.) Construct the section $a''b''a''b''$, showing the parts cut by the plane AB. Draw the line CD, which will be the elevation of a plane cutting the axis of the cylinder at an angle

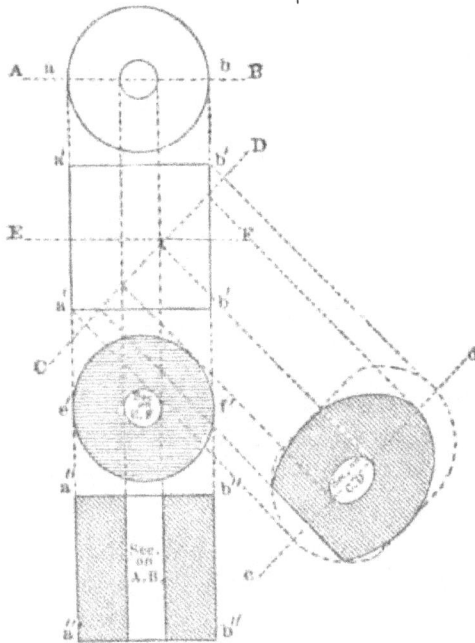

Fig. 128.

of 45°, and $\frac{3}{16}$ in. from the base. From the points where this line crosses the different lines of the elevation of the cylinder draw perpendicular leaders, and perpendicular to these draw an axis, cd. Lay off on these leaders on each side of the axis the widths of the cylinder at the points from which the leaders are drawn. Draw the outline of the section through these points, and shade that part that is cut by the plane. The several drawings so made will completely satisfy the requirements of the problem.

Prob. 2.—Make a drawing of a section taken through the axis of a cylinder 12 in. in diameter, 18 in. long, and 6 in. bore; scale, ¼ in. to the inch.

Prob. 3.—Make a drawing of a section taken 3¼ in. from the center and parallel to the axis of a cylinder 14 in. in diameter, 5 in. bore, and 6 in. long; scale, ¼ in. to the inch.

Prob. 4.—Make a drawing of a section of a cone, 9 ft. base and 1½ ft. pitch, pierced parallel to its axis and 3 ft. from it by a hole 2½ ft. in diameter, and cut by a plane passing through the center of the hole and the axis of the cone; scale, ½ in. to the foot.

Prob. 5.—Make a section drawing of a rectangular timber 2 ft. long, 3½ in. wide, and 9 in. thick, cut by a plane making an angle of 45° with the narrower edges; scale, 1 in. to the foot.

Prob. 6.—Make a section drawing of a cylinder 3 ft. in diameter and 6 ft. long, cut through its center by a plane making an angle of 45° with the axis; scale, 1 in. to the foot.

Prob. 7.—Make a section drawing of a hollow cylinder 21 in. in diameter, 42 in. long, and 7½ in. bore, cut 18 in. from one end by a plane making an angle of 45° with the axis; scale, 8 in. to the inch.

Prob. 8.—A cubical block 8 in. square is pierced, perpendicularly to its faces, by three holes 2 in. in diameter, passing through its center. Make a drawing of a section taken through two of its diameters. Make a drawing of another section taken through two of its diagonals; scale, ½ in. to the inch.

Prob. 9.—Make a drawing of a vertical section taken through the center of the school-room; scale, 4 ft. to the inch.

Prob. 10.—Draw a small two-story house in plan and elevation, and make a drawing of a vertical section taken through the center; scale, 8 ft. to the inch.

Prob. 11.—Draw a common pump, and make a vertical section; scale, ½ ft. to the inch.

Prob. 12.—Procure a common spool. Make a plan and elevation, and an oblique section drawing; scale, full size.

Prob. 13.—A right circular cone is 5 in. in diameter at the base, and 11 in. high. Make a drawing of a section, cutting the axis 5 in. from the top at an angle of 45°; scale, 2 in. to the inch.

CHAPTER X.—Foreshortened Dimensions.

A **Foreshortened Dimension** is a dimension that is parallel to neither its plan nor its elevation.

Prob. 1.—Make a constructive drawing of a triangular pyramid; sides of base, 3 ft., 2 ft., and 4 ft., and pitch 3 ft. 6 in.; plans of hips bisect the angles of the plan of the base; scale, ¼ in. to the foot.

SOLUTION.—Let *ABDC* (Fig. 129) represent the pyramid pictorially, and it will be seen that each hip is the hypothenuse of a right-angled triangle, of which the other two sides are the plan and the pitch. The problem, then, is to describe this triangle, and by so doing get

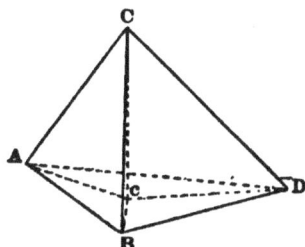

Fig. 129.

a line that represents the required hip to the given scale. Draw the plan and elevation of the pyramid *abcd* and *a'b'c'd'*, and at one end of the plan of each hip draw a perpendicular equal to the pitch. Join the end of this line with the other end of the plan of the hip. (Fig. 129, *a*). This line will be the hypothenuse of a right-angled triangle, of which the other sides are, respectively, the plan of the hip and the pitch of the pyramid. It, therefore,

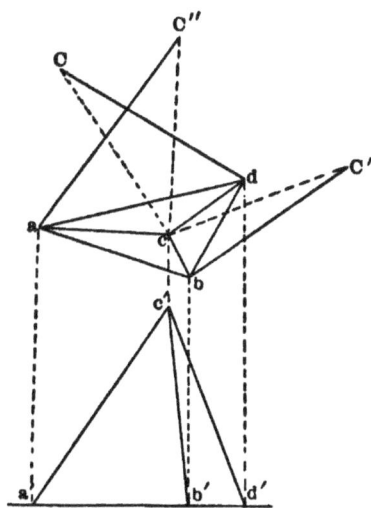

Fig. 129, *a.*

represents the true length of the hip, and *aC''*, *cC'*, and *dC* are the true lengths of the hips *AC*, *BC*, and *DC* represented to the scale of ½ in. to the foot.

Prob. 2.—Make a constructive drawing which will describe the true length of the hips of a pyramid 6 ft. square base, and 12 ft. pitch; scale, ¼ in. to the foot.

Prob. 3.—What is the length of the hip of a pyramid 9 ft. × 15 ft. base, pitch 12 ft., and ridge 6 ft.; scale, $\frac{1}{8}$ in. to the foot? (Fig. 130.)

Fig. 130.

Prob. 4.—Make constructive drawings for an oblique pyramid with base $4\frac{1}{2}$ ft. square, altitude 6 ft., and the plan of the apex 2 ft. outside of the base, and opposite the middle of one side; scale, $\frac{1}{2}$ in. to the foot.

Prob. 5.—Make constructive drawings for an oblique hexagonal frustum of a pyramid. Diameter of base, 21 in.; altitude, 27 in.; diameter of top, 13 in.; angle of axis with base, 60°; scale, $\frac{1}{8}$ in. to the inch.

Prob. 6.—Make constructive drawings for a framework 15 ft. square at base, 6 ft. square at top, 8 ft. high, outside measurement, and made of timbers 12 in. square; scale, $\frac{1}{4}$ in. to the foot.

Prob. 7.—Make a drawing showing method of framing timbers in frame-work described in problem 6; scale, 1 in. to the foot.

Prob. 8.—A heavy piece of machinery has fallen into a cistern 3 ft. 6 in. in diameter, that opens into a room 11 ft. high. It is required to construct the largest possible tripod crane with which to lift it out, the timbers of the crane to be 8 in. × 12 in., and the feet of the crane must not be placed nearer than 1 ft. from the opening of the cistern. Make working drawings for crane; scale, $\frac{1}{4}$ in. to the foot. (Fig. 131.)

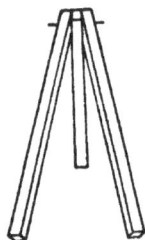

Fig. 131.

Rule.—Draw the plan and elevation of the given line. From one end of the plan draw a perpendicular equal to the pitch. Join the end of this line with the other end of the plan by a right line. This line will be the true length of the given line.

CHAPTER XI.—Development of Surfaces.

To **Develop the Surface** of any given object is to describe a diagram on some thin material, which, being cut out and bent or rolled into the proper shape, will inclose a space exactly equal and similar to that occupied by the given object.

Section I.—Development of Plane Surfaces.

Prob. **1.**—Develop the surface of a rectangular box $1\frac{1}{4}$ in. wide, $1\frac{3}{8}$ in. long, and $1\frac{3}{4}$ in. high.

Solution.—D r a w a plan of one side, *abcd*, Fig. 132, $1\frac{3}{8}$ in. long, and $1\frac{1}{4}$ in. wide. Draw the elevation of each of the adjacent sides *add″a″*, *aa′b′b*, *bb″c″c*, and *dcc′d′*. Draw the plan of the remaining side of the box, *d′c′b‴a‴*. Cut this diagram out and fold it on the appropriate lines,

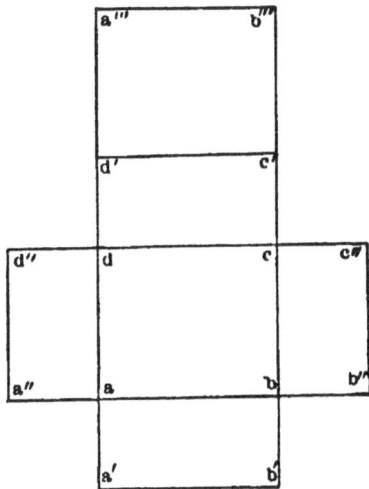

Fig. 132.

and it will exactly envelop a rectangular space $\frac{11}{16}$ in. $\times$ $1\frac{3}{8}$ in. $\times$ $\frac{13}{32}$ in. It is, therefore, the required developed surface of the given box.

Prob. 2.—Develop the surface of a rectangular box 2 in. $\times$ $1\frac{1}{2}$ in. $\times$ 1 in.

Prob. 3.—Develop the surface of a cube $\frac{3}{4}$ in. square.

Prob. 4.—Develop the surface of a square pyramid $1\frac{1}{2}$ in. base and 3 in. pitch.

Prob. 5.—A pan is $\frac{3}{4}$ in. deep, the bottom is $2\frac{1}{2}$ in. $\times$ 3 in., and the sides flare $\frac{1}{2}$ in. Develop the surface.

Prob. 6.—Develop the surface of a frustum of a rectangular pyramid 3 in. square at the base and 6 in. pitch; the plane of the top cutting the axis at an angle of 30°, 3 in. from the base.

NOTE.—Pitch, as used in this book, always means completed height; as, for instance, the pitch of a frustum is equal to the height of the completed object (cone or pyramid).

Prob. 7.—Develop the surface of a hexagonal prism 2 in. long and $\frac{3}{4}$ in. diameter.

Prob. 8.—Develop the surface of a hexagonal pyramid $1\frac{1}{4}$ in. diameter at the base and $\frac{5}{8}$ in. pitch.

Rule.—Draw the plan of any face of the given object; and, joining the sides of this, draw the faces adjacent to it. Draw the face or faces adjacent to these, and so on until all the faces have been described. The resulting diagram will be the required developed surface.

Section II.—Development of single curved surfaces.

A **Single Curved Surface** is a surface through any point of which one straight line can be drawn that will lie wholly within the surface. Cylindrical and conical surfaces are of single curvature.

Cylindrical Surfaces.

Prob. 1.—Develop the surface of a cylinder ½ in. long and ⁵⁄₁₆ in. in diameter.

SOLUTION.—Suppose *AB* (Fig. 133) to be the given cylinder, and imagine that it has been freshly painted and allowed to roll round once on some flat surface. The print of the paint on this surface would describe the developed curved surface

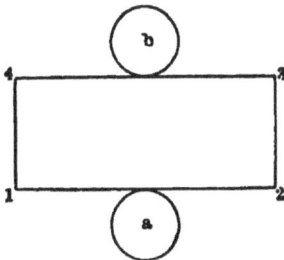

Fig. 133.

of the cylinder. The length of this print mark would be equal to the circumference of the cylinder and the width would be equal to the length of the

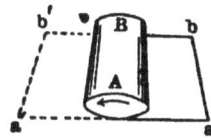

Fig. 134.

cylinder. The circumference of the cylinder given in this problem is equal to ⁵⁄₁₆ in. × 3.1416 = .9812 in. = 1 in., and the width is ½ in. Draw the rectangle 1, 2, 3, 4, 1 in. long and ½ in. wide, and it will be the developed curved surface of the cylinder. (Fig. 134.)

Draw two circles ⁵⁄₁₆ in. in diameter, tangent to the

sides of this rectangle, and the development is complete.

Prob. 2.—Develop the surface of a cylinder 1½ in. long and ¾ in. in diameter.

Prob. 3.—Develop the surface of a cylinder 3 in. long and ⅜ in. in diameter.

Prob. 4.—Develop the surface of a cylinder 1¼ in. long and ⅛ in. in diameter.

Prob. 5.—A hollow cylinder, 1¾ in. in diameter outside, and ⅞ in. in diameter inside, and 1½ in. long, is to be made of tin. Develop the surface inside and outside as well as at the ends.

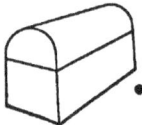

Fig. 135.

Prob. 6.—Develop the surface of a sheet-iron box that is ¼ of a cylinder 2 in. in diameter and 2 in. long.

Prob. 7.—Develop the surface of a paper box 3 in. long, 2 in. wide, and 1 in. high, with a half round cylindrical cover. (Fig. 135.)

Section III.—Cylinders, the ends of which are not parallel.

Prob. 1.—Develop the surface of a cylinder ⅞ in. in diameter, ¾ in. long at the axis, and cut off at one end at an angle of 45°.

SOLUTION.—Draw the plan, 1–12, ⅞ in. in diameter, and the elevation, 1'-7', ¾ in. long at the axis, and cut off at the top at an angle of 45°. (Fig. 136.)

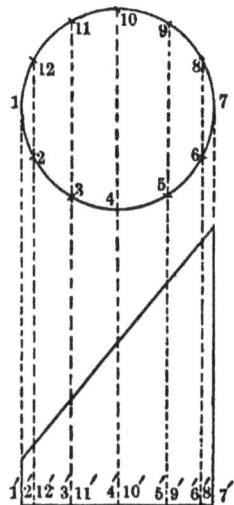

Fig. 136.

Divide the plan into 12 equal parts. Draw leaders from these division points, and find the elevations of the corresponding elements of the cylinder, 2′-7′.

Draw a line *AB*, equal in length to the circumference of the cylinder, 1 in. × 3.1416 = 3.1416 in. = $8\frac{1}{64}$ in. Divide this line into 12 equal parts, and at the division points erect perpendiculars equal in length to the elevations of the corresponding ele-

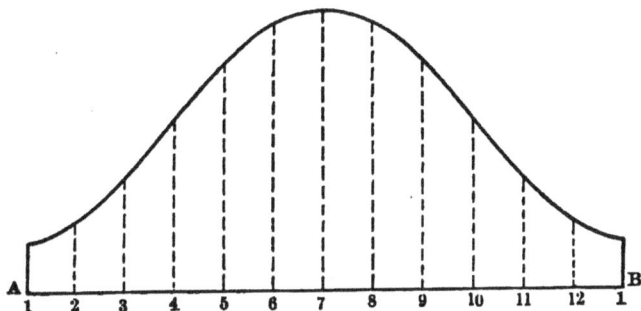

Fig. 137.

ments of the cylinder. Join the tops of these lines, and the resulting diagram is the required developed surface of the given cylinder. (Fig. 137.)

Prob. 2.—Develop the surface of one half a rectangular elbow of a round sheet-iron pipe, $2\frac{1}{2}$ in. long in the longest element, and 1 in. in diameter.

Prob. 3.—Develop the surface of a $1\frac{1}{2}$ in. cylinder, cut off 1 in. from the base at an angle of 30° with the axis.

Prob. 4.—Develop the surface of a $1\frac{3}{4}$ in. cylinder, $1\frac{1}{4}$ in. long, one end making an angle of 60° with the axis.

Prob. 5.—Develop the surface of one half of a 60°
elbow, ¾ in. diameter and 1⅛ in. at its longest element.

Prob. 6.—Develop the surface of one half of a
150° elbow, ⅞ in. in diameter, and ⅝ in. at its
shortest element.

Section IV.—Development of conical surfaces.

Prob. 1.—Develop the surface of a cone; base, 1⅛
in. in diameter and pitch ${}_{3}^{1}{}_{2}$ in.

SOLUTION.—Let *bAc* be the given cone lying on

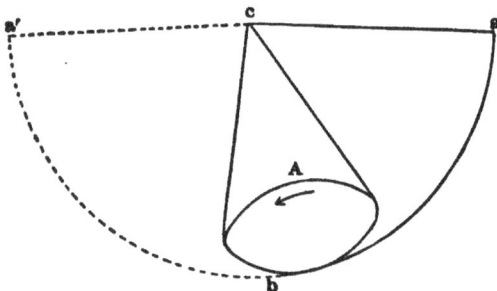
Fig. 138.

its side and in
the process of
revolving. At
the beginning
of the revolu-
tion *A* was at *a*,
and at the end
A will be at *a'*.
Then the length
of the arc *a'ba* will be exactly equal to the circumference of the circle *Ab*, at the base of the
cone, and *ca*, the radius with which it has been described, is exactly equal to the slant height of the
cone. (Fig. 138.)

Draw the plan and elevation of the cone, *abc* and
a'b'c' (Fig. 139), having the dimensions given in the
problem. From this ascertain the slant height, *a'c'*,
and with $AC = a'c'$ as a radius, describe the arc of
a circle, *ABA'*. (Fig. 140.)

The number of degrees that are to be laid off on
this circle in order that the length of the arc

shall be equal to the circumference of the base of the cone will bear the same relation to 360° as the radius of the base of the cone bears to its slant height.

To ascertain this, divide the slant height of the cone $1\frac{1}{8}$ in. by the radius of the base of the cone $\frac{9}{16}$ in. $\dfrac{1\frac{1}{8}}{\frac{9}{16}} = \dfrac{\frac{9}{8}}{\frac{9}{16}} = \frac{8}{16} = \frac{1}{2} = 180°.$ Lay off on the arc ABA' 180°, and draw the radii AC and $A'C$. The inclosed sector is the developed surface of the given cone.

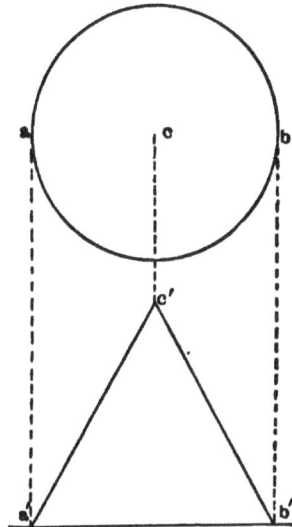

Fig. 139.

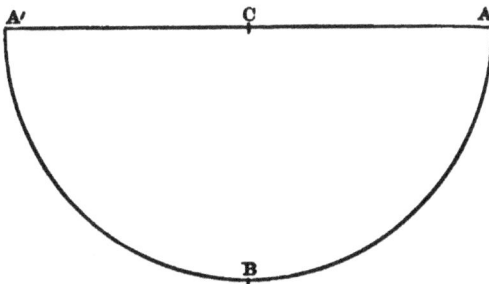

Fig. 140.

Prob. 2.—Develop the surface of a cone $1\frac{1}{2}$ in. base and $1\frac{3}{4}$ in. pitch.

Prob. 3.—Develop the surface of a cone 2 in. base and $\frac{1}{2}$ in. pitch.

Prob. 4.—Develop the surface of a cone $\frac{1}{2}$ in. base and $2\frac{1}{2}$ in. pitch.

Prob. 5.—Develop the surface of a frame 1 in. wide and $1\frac{1}{4}$ in. pitch; 2 in. long and half conical at the ends. (Fig. 141.)

Fig. 141.

Rule.—Draw plan and elevation of cone, and describe a circle having a radius equal to the slant height of the cone. Lay off on this circle an arc containing a number of degrees that have the same ratio to 360° that the radius of the base has to the slant height of the cone. Draw radii from the center to the ends of this arc, and the sector thus inclosed will be the developed surface of the cone.

Section V.—To develop the surface of the frustum of a cone.

Prob. 1.—The frustum of a cone has a base 1 in. in diameter; the length of the axis is $\frac{15}{32}$ in.; the angle at the base is 70°, and the top is cut off at an angle of 45° with the axis. Describe the developed surface of the frustum.

SOLUTION.—Draw the plan and elevation, *abc* and

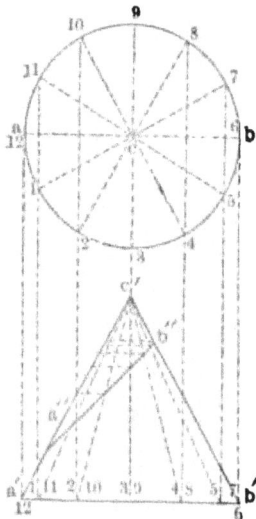

Fig. 142.

a'b'c', and describe the arc *ABC*, the length of which is equal to the circumference of the circle *ab*. Divide the plan into any number of equal parts, and draw radial lines to the center *c*. Draw elevations of these lines in *c'*1-7, *c'*2-6, and *c'*3-5. Divide *ABA'* into the same number of equal parts as the plan *ab*, and draw *C*1, *C*2—*C*7. From the points where the elevations of the elements meet the line *a''b''*, the top of the frustum, draw horizontal lines to meet *c'b'* in

a'', 1-7, 2-6, and 3-5. From C lay off the distances
$c'a''$, $c'1$, $c'2$—$c'7$, $c'a''$. Join the points just found
by a curved line, $a''1234567a''$ and $ABA'a''a''$, the

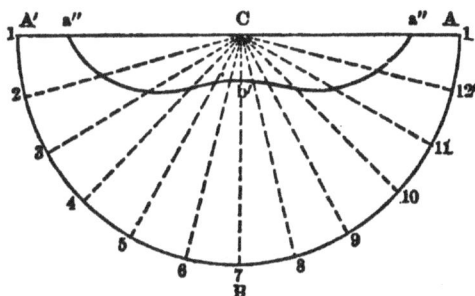

Fig. 143.

inclosed figure will be the developed surface of the
frustum described in the problem. (Figs. 142 and
143.)

Prob. 2.—Develop the surface of the frustum of a
cone 1¾ in. base, 1⅛ in. axis, 75° angle at the base,
60° angle of top with axis.

Prob. 3.—Develop the surface of the frustum of a
cone 1¼ in. base, ¾ in. axis, 30° angle at the base,
40° angle of top with axis.

Prob. 4.—Develop the surface of the frustum of a
cone ½ in. base, 2 in. axis, 75° angle at base, 90°
angle of top with element of cone.

Prob. 5.—Develop the surface of the frustum of a
cone 2 in. base, ½ in. axis, 22½° angle at the base,
45° angle of top with element of cone.

Section VI.—Development of surfaces of double curvature.

Prob. 1.—Develop the surface of a sphere 1 in. in
diameter.

SOLUTION.—Draw the plan and elevation of the sphere, *abc* and *a'b'c'*. (Fig. 144.) Divide one half of the circumference of the elevation into any number of equal divisions at the points *II*, *III*, *IV*, and through these points draw the elevations of horizontal circles. Find the plans of these circles. Divide the plan, *abc*, into any number of equal sectors, 1, 2—5, 6. Find the elevations of the points of intersection of the sides of the section 1-2 with the plans of the horizontal circles. Draw a line, the length of which is equal to one half the circumference of the sphere, *AB*, 1.57 in. Divide this line to correspond with the divisions of the elevation of the sphere, and through these points draw perpendicular lines equal to the arcs of the circles at *II*, *III*, *IV*, *V*, *VI*, in the elevation. Join the ends of these lines by a curve, and the inclosed space will be the developed surface of one sixth of the whole surface of the sphere, which, being reproduced six times, will completely envelop the sphere.

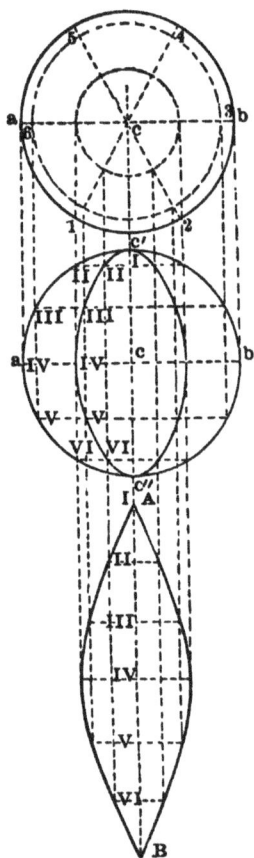

Fig. 144.

Prob. 2.—Develop the surface of a sphere 1¾ in. in diameter.

Prob. 3.—Develop the surface of a sphere 2⅛ in. in diameter.

Prob. 4.—Develop the surface of a sphere ½ in. in radius.

Prob. 5.—Develop the surface of a sphere 1⅝ in. radius.

Rule.—Draw plan and elevation of the sphere. Divide the circumference into any number of equal divisions. Through these division points draw elevations of horizontal circles. Draw the plans of these circles. Divide the plan of the sphere into any number of equal sectors, and find the elevation of that sector that is parallel to the elevation. Draw a line equal in length to one half the circumference of the elevation, divide it to correspond with the elevation, and through the division points draw perpendicular lines, the lengths of which are equal to the width of the sectors at the different points. Join the ends of these perpendiculars, and the inclosed figure will be the developed surface of one section of the given surface.

CHAPTER XII.—Development of Oblique Conical Surfaces.

An **Oblique Cone** is a cone in which the axis is not perpendicular to the base. (Fig. 145.)

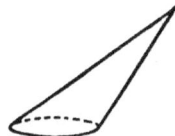

Fig. 145.

Prob. 1.—Develop the surface of an oblique cone; base, 1¼ in. in diameter, 1⅛ in. pitch, and axis 45° with the base.

SOLUTION.—Draw a plan and elevation of the cone, and divide the plan of the base into any number of equal parts. Draw the plans and elevations of the elements of the cone at these points. These elements divide the surface of the cone into small triangles, the bases of which are equal portions of the base of the cone, and the sides are foreshortened dimensions. (Chap. X.)

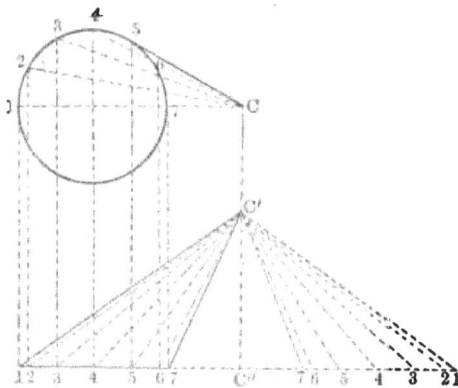

Fig. 146.

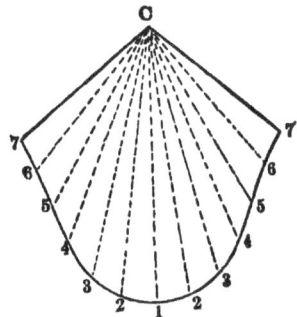

Fig. 147.

To ascertain the length of the foreshortened sides of these triangles, draw the pitch line $C'C''$ and extend the base of the cone beyond it. Lay off on this extended base from c'' the lengths of the plans of the elements of cone, $c''1$, $c''2$-$c''7$. Join $c'1$, $c'2$-$c'7$, and these lines are the true lengths of the foreshortened sides of the required triangles. (Fig. 146.)

Now draw a line $C1$ equal to $c'1$, and with the end 1 as a center, describe arcs on both sides of $C1$, of which the radius is equal to $\frac{1}{12}$ the circumference of the base of the cone. With C as a center, and a radius equal to $C''2$, describe arcs cutting these. There are now described two adjacent triangles on the surface of the cone. In the same manner describe the triangles adjacent to these until all have been described. (Fig. 147.)

Prob. 2.—Develop the surface of a cone of which the base is 1⅜ in. in diameter, pitch 1⅛ in., and axis makes an angle of 60° with base.

Prob. 3.—Develop the surface of an oblique cone; base, ⅝ in. radius; pitch, ¾ in.; and 22½° axis with base.

Prob. 4.—Develop the surface of a frustum of an oblique cone; base, 1 in. radius; pitch, 2 in.; axis, 1 in. long, 45° with base.

Prob. 5.—The top of a portable furnace 2 ft. 6 in. in diameter is 2 ft. 6 in. below the under side of the floor joists, and it is required to construct a conical top which will connect by a straight pipe 6 in. long with a register box 12½ in. × 19½ in., of which the center is 13 in. outside of the furnace top, and the longer edges make an angle of 30° with the axis of the frustum. Develop the surface of the frustum.